How to Copyedit Scientific Books & Journals

The Professional Editing and Publishing Series

This volume is one of a series published by ISI Press®. The Professional Editing and Publishing Series provides timely, practical information to help publishers and editors of professional, scholarly, and scientific publications.

Books published in this series:

How to Edit a Scientific Journal
    by CLAUDE T. BISHOP

*Copywriter's Handbook: A Practical Guide for Advertising and
    Promotion of Specialized and Scholarly Books and Journals*
    by NAT G. BODIAN

An Insider's Guide for Medical Authors and Editors
    by PETER MORGAN

How to Copyedit Scientific Books and Journals
    by MAEVE O'CONNOR

Books to be published in this series:

*Medical Style and Format: An International Manual for
    Authors, Editors, and Publishers*
    by EDWARD J. HUTH

How to Copyedit Scientific Books & Journals

Maeve O'Connor

iSi PRESS®

Philadelphia

Published by

iSi PRESS®   A Subsidiary of the
Institute for Scientific Information®
3501 Market St., Philadelphia, PA 19104 U.S.A.

© 1986 Maeve O'Connor

Library of Congress Cataloging-in-Publication Data

O'Connor, Maeve.
  How to copyedit scientific books & journals.

  (The Professional editing and publishing series)
  Bibliography: p.
  Includes index.
  1. Science publishing—Handbooks, manuals, etc.
2. Scientific literature—Editing—Handbooks, manuals,
etc. 3. Copy-editing—Handbooks, manuals, etc. I. Title.
II. Series.
Z286.S4028 1986   070.5   86-20073
ISBN 0-89495-070-3
ISBN 0-89495-064-9 (pbk.)

Printed in the United States of America.
93 92 91 90 88 87 86   8 7 6 5 4 3 2 1

Contents

vi   Contents

Preface

So you want to be a copyeditor, or get into publishing, and copyediting in science or medicine seems a good way to start? "After all," you say, "it's just correcting the grammar, spelling, and punctuation, isn't it?" Well, yes, that's part of it — but it isn't quite as simple as that description sounds. Copyediting, especially in science (including medicine), is complex and demanding. Copyeditors in science need some appreciation or knowledge of science as well as of grammar. They should have patience, common sense, a passion for accuracy, and the ability to compromise and to concentrate. They should own a good pair of eyes, and a strong back and seat. And they should possess a resilient nature — for coping with authors' frailties or their own, and for withstanding the harassments created by production deadlines.

A copyeditor's life isn't all harassment: those who survive the first months or years find the hidden rewards. There is the sense of accomplishment when you have reshaped a long-winded paragraph into a concisely written one, or converted an opaque sentence into a transparent one while keeping the author's meaning. There is the feeling you may have helped science forward a millimeter or so when you have put the references to rights, caught errors in tables, and cropped photographs to highlight the important parts. And there is the satisfaction of a job well done when a somewhat imperfect manuscript has been transformed into an article acceptable to the editor, the typesetter, the reader — and, of course, the author.

When I was asked to write a book on how to achieve this kind of transformation, I said that the do's and don't's of copyediting had already been spelled out by Judith Butcher (*Copy-editing: The Cambridge Handbook*) and Karen Judd (*Copyediting: A Practical Guide*). Then I realized that Bob Day, then Director of ISI Press, was (of course) right. Those two splendid books do not deal specifically with copyediting scientific journals or books and nothing else fills the gap. Yet copyeditors in science need special help.

After all, they have to deal with difficult or esoteric subjects and they usually work for editors whose true profession is science, not editing — leaving extra scope and responsibilities for their copyeditors.

Courses for copyeditors in science, like books on the subject, are scarce, and most people learn copyediting by doing it, sometimes without anyone sitting beside them. New copyeditors or even well-established freelances may therefore be unsure whether they are doing the right thing or working in the best or most efficient way. This book aims to explain the job to new or would-be copyeditors, including freelance workers and authors' editors, in science or medicine. It should also interest editors themselves — some do their own copyediting and the rest may want to know what their copyeditors are up to.

The first chapter defines copyediting and says something about why it is needed, who is qualified to do it, and in which order its various aspects should be tackled. The second chapter describes how manuscripts are handled and marked up in journal offices (but some of this chapter applies to books). The heart of the book lies in the chapters on substantive and technical editing — these, too, refer particularly to scientific journals but are also intended for copyeditors working on books. Then there is more on mark-up and coding electronic "manuscripts" for typesetting or other methods of transferring information, followed by chapters on completing journal issues, proofreading, and the special problems of copyediting conference proceedings, symposia, and other scientific books. The book ends with lists of useful addresses and further reading.

Read the book all the way through, if you can, and afterwards use it as a reference book. It is designed to be a general guide to copyediting in science, for use with the books by Butcher or Judd already mentioned and with the style manuals referred to later. It supplements these bibles but in no way supplants them.

Acknowledgments

I am very grateful indeed to Sue Burkhart, Sarah Clark, Frances Porcher, Jane Smith, and Julie Whelan for commenting on the manuscript at different stages and providing encouragement along the way. I also want to thank Bob Day, Maryanne Soper, and the anonymous reviewers who commented critically but — on the whole — supportively on the proposal and the draft manuscript. Many others have contributed help, wittingly or unwittingly, and my thanks go to them too, if they should read this book. My thanks and sympathy go to Estella Bradley, who copyedited the manuscript. My other debts will be clear from the preface and the references in the text. Omissions, oddities, and errors are all my own work.

Chapter 1

Copyediting Construed

Copyediting, like editing itself, is a mystery to most people. For one thing, copyeditors are hidden behind a score of aliases — subeditor, technical editor, developmental editor, desk editor, manuscript editor, author's editor, line editor, redactor, and editorial assistant among them. For another, the job is complex and difficult to explain to anyone outside the publishing world.

What Do Copyeditors in Science Do?

Whatever their job titles, copyeditors in science do several types of work in several stages (see Fig. 1). In one type of work, here described as administration, they may direct the traffic of manuscripts between the editor, the reviewers or referees, the authors, and the typesetter and printer. In a second type, substantive editing, copyeditors may reorganize material and even rewrite some of it. In a third type, technical editing, they correct grammar, punctuation, and spelling; mark mechanical style; and code manuscripts or mark them up for typesetting. (See Figs. 2 to 4.)

Administration

The first kind of work is not copyediting, strictly speaking, but copyeditors often deal with the organizational tasks listed in Fig. 2. They may also cope with copyrights and permissions, write to authors and reviewers, persuade typesetters and printers to keep to their promised schedules, and mail proofs to authors.

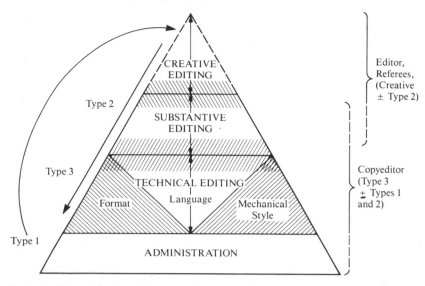

Figure 1   Types of copyediting, showing areas of overlap between creative and substantive editing, and between substantive editing and technical editing (see Figs. 2 to 4 and text).

---

(1) On receipt of manuscript
Assign reference number
Record date of receipt, authors, title
Check that all items required are present
Check that the text and other sections are typed in double spacing (for preference)
Write the reference number on title page and elsewhere
Send acknowledgment card

(2) On acceptance
Check whether releases are needed for borrowed material
Send acceptance letter and copyright form
Check that copyright form is returned

(3) During production process
Check progress of manuscript
Deal with proofs and orders for reprints

Figure 2   Copyediting type 1: administration. Copyeditors working in publishing houses may have additional responsibilities (see Chapter 10). This figure and the next two, and several chapters in the book, owe much of their structure to Robert Van Buren and Mary Fran Buehler's invaluable booklet, *The Levels of Edit*.[1]

Make titles and abstracts concise, accurate, and informative
Add or obtain key words and running heads (or footlines), if needed
Check sections and their headings
Examine the logic, order, correctness, and emphasis of presentation
Eliminate repetition, redundancy, irrelevancies
Suggest ways of shortening words, phrases, sentences, paragraphs
Examine tables in relation to text
Make table titles appropriate, and parallel if in a series
Check presentation of units in column headings and numbers in body of tables
Examine figures in relation to text
Check suitability of figures for reproduction
Make figure legends appropriate, and parallel if in a series
Crop or mask photographs to emphasize important parts
Delete excessive detail from line drawings

Figure 3 Copyediting type 2: substantive editing.

Substantive Editing

On another level some copyeditors are responsible for the substance of manuscripts. That is, they may be expected to rewrite titles and abstracts of articles or chapters to make them concise yet informative, and they may have to realign authors' logic, rewrite parts of the text, restructure tables, reshape illustrations, and query statistics and equations (Fig. 3). This is *substantive editing*, defined by the *Chicago Manual of Style*[2] (p. 51) as "rewriting, reorganizing, or suggesting other ways to present material."

Substantive editing means suggesting minor yet important ways in which papers might be rewritten or reorganized for greater readability and precision. It may be done heavily, moderately, lightly, or not at all, depending on the state of the manuscript, the policy of the publication, the time available, and how much the copyeditor knows about the subject. The rewriting aspect overlaps with technical editing on the one hand and creative editing on the other. *Creative editing* consists, in part, of suggesting or requesting major changes in the content or organization of a manuscript; this kind of editing is the preserve not usually of the copyeditor but of the editor and the reviewers or referees who advise the editor. In practice, though, copyeditors often act as "the referee of last resort," as Robert Schoenfeld[3] puts it.

Technical Editing

The third type of work done by copyeditors is usually called *mechanical* or *technical* editing (but "technical editing" is sometimes used to refer

---

(1) Correct the language
Spelling
Punctuation
Grammar and syntax
Usage

(2) Deal with mechanical style (style related to content) in:
Spelling (consistency); hyphenation and compound words
Capitals, italics, bold face
Abbreviations, acronyms, symbols
Non-roman alphabets, special scripts, and accents
Numerals
Parentheses, brackets, mathematical braces
Spacing between or around characters
Tables
Figures, equations, structures, maps
Linking symbols for curves or footnotes

(3) Indicate the format (visual style) for:
Type faces, type sizes and leading
Column/page width
Title page (chapter headings)
Abstract or summary
Key words and bibliographic identifiers
Headings
Indention
General layout (size, shape, and position of tables and figures)
References
Form and position of lettering on figures

Figure 4   Copyediting type 3: technical editing.

to the whole process of editing technical material; the terminology here is fluid) (Fig. 4). The *Chicago Manual of Style*[2] (p. 51) defines this kind of editing as involving "a close reading of the manuscript with an eye to such matters as consistency of capitalization, spelling, and hyphenation; agreement of verbs and subjects; beginning and ending quotation marks and parentheses; number of ellipsis points; numbers given as figures or written out; and many other details of style." Technical editing includes correcting the grammar, ranking headings and subheadings, checking abbreviations and other conventions in particular subjects, styling references, and checking that all the necessary bibliographic information is present in the reference list or in footnotes.

Marking up text for the human typesetter, coding printouts from floppy disks or magnetic tapes ("compuscripts") for computer typesetting direct from these electronic media, and checking that author-prepared camera-ready copy is indeed ready to go before the camera lens are other essential parts of technical copyediting that come under the heading of format.

Copyediting in Science: A Definition

If copyediting in science can be defined at all, it consists of correcting the language, marking the mechanical style and format of scientific manuscripts to bring them to the required standards for publication, and coping, if required, with substantive editing, administration, and proofreading — or perhaps doing only some or none of the last three. Copyediting puts manuscripts under a microscope; creative editing deals with them at the macroscopic level — getting the science right. Copyediting is in fact "the editor's most important and most time-consuming task," as the *Chicago Manual of Style* puts it[2] (p. 51).

What Are Copyeditors Employed to Do?

The overall responsibilities of copyeditors employed by book or journal publishers are to prepare manuscripts for publication and, in the process, help editors to strike a balance between the interests of authors and those of readers. Authors who have had an article accepted by a journal or for a multi-author book are mainly interested in seeing it in print quickly, with no further damage to their egos or demands on their time. Readers, however, want articles that are quick to read, with everything this implies; they want useful and comprehensible tables and figures; and they need useful and accurate reference lists which represent the author's building-blocks fairly.

One kind of copyeditor has somewhat different responsibilities: *authors' editors* employed by research departments of various kinds, or by individuals, owe their first duty to authors when manuscripts are being prepared for submission rather than to editors after manuscripts have been accepted for publication. Authors' editors nearly always do substantive editing, and sometimes creative editing or writing, as well as technical editing. They may also advise authors on where to submit their manuscripts and what to do if their work is rejected or if major revision is requested.[4]

Another important responsibility of copyeditors is to guard their employers' interests by keeping production costs down — for example by preparing manuscripts carefully and keeping proof corrections to a minimum.

Why Is Copyediting Needed?

Some authors scarcely realize that copyediting exists. Others denounce it as an unwanted luxury or as an exercise in pedantry, obfuscation, or "promiscuous depredation," as Barzun put it[5] — performed by people who edit style out and "delight in trying to turn straightforward English into jargon."[6] Copyeditors have even (jokingly?) been called murderers: "Foul fiend, you've killed my manuscript/with bodkin pencil bloody tipped."[7] Editors new to the job may wonder why papers they and their referees have accepted for publication can't go straight to the typesetter without the intervention of a copyeditor. Authors may ask why anyone should be paid to alter -ise endings to -ize or -or to -our, or vice versa, or to ruin their meaning by changing carefully considered wording or punctuation. "Just change the 'was's' and 'were's'," those authors say.[8]

So why *are* copyeditors a necessary part of the publishing process? Scholarly journals and books, especially in science, aim to publish work that is correct and consistent, concise and precise, in every detail of form and substance, as well as being important, true, and comprehensible.[9] If readers find mistakes in details, they are likely to be suspicious of the rest of what the author has to say. In science correctness in seemingly minor matters is all-important, because readers may want to replicate experiments, observations, or practical applications. Since most scientific authors are scientists first and careful writers a long way afterwards, copyeditors are needed in scientific and technical publishing even more than they are in general publishing.

In dealing with correctness, consistency, precision, and comprehensibility, copyeditors save publishers' money, readers' time and tempers, and authors' reputations. They keep production costs down by styling manuscripts in an agreed way, cutting out empty phrases ("in this connection we can say that"), and making authors' intentions plain to typesetters or other keyboarders — who need to have every detail clearly marked if text is to be keyed in quickly. They sift out minor errors of a kind that annoy readers and are easily overlooked by editors and referees. They detect columns in tables that don't add up correctly and dates in the text that clash with those in the reference list. And they pounce on words that are spelled or typed wrongly ("recumbent DNA" was one that got away . . .).[10]

Copyeditors do far more than change the was's and the were's to make the meaning clear. Their work is particularly heroic in science, where most people are more interested in getting on with their next piece of work than in writing up the last set of experiments, and where editors are often working scientists who have little time for the detailed work of preparing manuscripts for publication. In science, as elsewhere, it is copyeditors who defend readers from the worst that authors can do, while protecting authors from their own mistakes.

How Can Copyediting Be Justified?

Copyediting seems to be needed; can it be justified? Is it honest to alter other people's way of writing, making authors appear better than they are? Shouldn't writers improve their own prose and clarify their meaning themselves, if an editor asks them to do so? In science, the problem here is the average author's dislike of writing and lack of training in it. The written language is almost an alien one to such authors.

Far from editing style out, editors and copyeditors are often the only people who can make scientific papers moderately readable. In the humanities the way an article or book is written may be an important part of the message, so style should not be tampered with much, if at all. In science the message is everything; if the style of the language is unsatisfactory, it can legitimately be remodeled for the sake of getting the message over to readers. And because not even the best copyeditor can disguise bad science, copyediting as usually practiced in scientific publishing can hardly be described as deception.

What Qualities and Qualifications Do Copyeditors Need?

Scientific copyeditors should ideally possess "the patience to read over the same copy several times" and the "power to concentrate on the form and meaning of the text simultaneously," not to mention owning "strong eyes, back and seat."[11] They also need tact, modesty, willingness to compromise, and the ability to take "real pleasure in bringing order out of chaos" and "to spot quickly errors, inconsistencies, redundancies and ambiguities."[12] Other essential qualities are a passion for accuracy — but a well-controlled one; a sound knowledge of grammar; a feeling for language; a grasp of logic; "the wit to know when to stop comma-catching and let the book or journal get to press"[13] (p. 151; see also DeVivo[14]); and natural resilience, a good sense of humor, and common sense. An early love of words and books or any form of print is a promising background to have. The ability to write well is useful too, but it must go with the strength of mind to let authors say things their own way, provided they say them correctly — hence the First Edict of Copyediting:

> LEAVE WELL ENOUGH ALONE

Copyeditors working on scientific journals or books usually need some formal qualifications too. But there is no clear path to a career in copyediting in science: employers have different requirements according to the kinds of journals or books they produce and their experience, good or bad, with previous copyeditors. For most jobs the would-be copyeditor needs at least a bachelor's degree in either science or the humanities. It is true that the more basic the research being published, the more likely it is that manuscripts will range from the enigmatic to the impenetrable, and the more daunting the job will be for those lacking some research experience in the subject area. On the other hand, a degree in English "is not an absolute contraindication to an editing career in science,"[15] and some employers even prefer graduates in English or other non-scientific disciplines. Such employers argue that the vocabulary of a scientific discipline and something of its meaning can be assimilated, given time, but the ability to handle language sensibly and sensitively cannot. A sound knowledge and love of language and some experience in an appropriate branch of science will give you a head start. But if you have no degree, or no degree higher than a first one in a non-scientific subject, don't despair. Persevere, be patient, do some studying on your own, and try to get some training.

What Training Is Available for Copyeditors?

Most professional training in copyediting is provided by working at the job. Open courses in copyediting are rare, in scientific copyediting rarer still. If you can track a course down, invest in it, or persuade your employers to do so (see the appendix for some useful addresses). There are coursebooks — for example *Editing for Everyone* by Celia Hall[16] — which provide absolute beginners with a simple introduction to non-scientific copyediting that can be worked through at home. On another level are reference books such as those named below under "Tools for Copyediting."

In your first job, or first months at a new job, you may be reading proofs and checking references rather than editing text. This may be the only "training" you get, and you will be expected to assimilate the house methods and style as you work. As well as studying whatever reference books and manuals are available, note the kind of changes that have been made on the manuscripts you see, discover for yourself why they have been made, and ask questions about anything you don't understand.

Tools for Copyediting

Reference books of various kinds are the copyeditor's main working tools (after pens with different colored inks or with erasable ink, and pencils, erasers, and pencil-sharpeners). You or your employers should be prepared to spend a lot on reference books: they cost slightly less than gold but are well worth their weight in any currency.

You will need a good general dictionary (unabridged); a specialized dictionary or dictionaries for the disciplines you will be editing; the *Chicago*

Manual of Style[2] or Butcher[17] or Judd[18] on copyediting; the *CBE Style Manual*[19] or *Geowriting*[20] or a similar guide in other branches of science; Follett[21] or Fowler[22] and Strunk and White.[23] A standard grammar book[24,25] and a couple of the many books that advise scientists on how to write well would be useful additions to your bookshelf (see the list of further reading at the end of this book). Edward Huth's *Medical Style and Format*[26] will probably be extremely useful too, even if you are working outside medicine.

If you work for a journal, its style manual or the manual of the main journal in the discipline (or both) must be your constant companion(s). If no such manuals exist, the journal's instructions to authors will show how authors are expected to prepare their manuscripts and therefore what the copyeditor has to look for.

If house style (see p. 19) requires abbreviated titles of journals to be used in reference lists, you'll need a recognized list of title-word abbreviations or a list of journal names showing their approved abbreviations (for example, *Serial Sources for the BIOSIS Data Base*[27] or the list of journals indexed in *Index Medicus*[28]).

Sources for checking nomenclature and terminology in the specific field you are dealing with are essential. The *CBE Style Manual*[19] gives lists of sources in the life sciences. For units of measurement and their abbreviations a list of Système International (SI) units that includes conversion factors for older units is also essential (see Chapter 5 for more on SI units). Other publications you'll need to refer to, such as *Ulrich's International Periodicals Directory*,[29] *Current Contents*,[30] and *Science Citation Index*,[31] can be consulted in a library, as can textbooks on the subjects covered by the journal.

Other essential tools are a copy of the house style and your own style sheets, as described in Chapter 2.

This armory, with plenty of desk space, good lighting, and a comfortable chair will help you to tackle manuscripts successfully.

Professional Associations for Copyeditors (and Editors)

Another source of help for copyeditors is an editors' association. Anyone already working as a copyeditor will probably be accepted for membership of one of these. If you are not eligible for membership, the association nearest to you geographically or by scientific discipline may produce a newsletter to which non-members can subscribe, and its conferences or seminars may be open to non-members. In the United States and Canada, for example, the Council of Biology Editors (CBE) and the Association of Earth Science Editors (AESE) meet annually, and both publish newsletters (*CBE Views; Blueline*). The American Medical Writers Association (AMWA) holds short courses in scientific editing and writing, and publishes *Medical Communications*. In Europe, the European Association of Science Editors (EASE) accepts members from anywhere in the world and publishes a useful bulletin (*European Science Editing*). EASE, formed by an amalgamation between the European Life Science Editors (ELSE) and the European Earth Science Editors' Association (Editerra), holds its main conference and assembly every three years, with other meetings in between.

Other regions and countries also have associations for editors (for addresses see the appendix). The International Federation of Scientific Editors' Associations (IFSEA), which provides a forum for all of them, is mainly an association of associations but accepts individual members too. It holds an international conference for scientific editors every three years. One of the aims of IFSEA is "to contribute to the formulation of standards and good practices and to promote their international implementation," which it does by promoting courses in editing and copyediting in developing countries and by encouraging the formation of new associations of editors in areas where these are needed.

References

1. Van Buren R, Buehler MF.   1980   The levels of edit, 2nd ed. Jet Propulsion Laboratory, California Institute of Technology, Pasadena, CA (JPL Publication 80-1), 1980.
2. University of Chicago Press.   1982   The Chicago manual of style, 13th ed. University of Chicago Press, Chicago, 1982.
3. Schoenfeld R.   1982   Dollars-and-cents value of efficient presentation. IEEE Transactions on Professional Communication 1982; PC-25:144–150 [reprinted from the Journal of Chemical Information and Computer Sciences 1981; 21:61–66].
4. Gilbert JR, Wright CN, Amberson JI, Thompson AL.   1984   Profile of the author's editor: findings from a national survey. CBE Views 1984; 7(1):4–10.
5. Barzun J.   1985   Behind the blue pencil: censorship or creeping creativity? Publishers Weekly 1985; p. 28–30 (6 Sept).
6. Pirie NW.   1975   English and editorial boards [brief letter]. Nature 1975; 256:161–162.
7. Cherry K.   1972   That uncommon comma; or, all in the editor's morning mail. Scholarly Publishing 1972; 4:29.
8. Glen HW.   1981   User attitudes to scientific editing: just change the 'was's' and 'were's.' Journal of Research Communication Studies 1981; 3:221–227.
9. DeBakey L.   1976   The scientific journal: editorial policies and practices — guidelines for editors, reviewers and authors. Mosby, St Louis, MO, 1976.
10. Advertisement in *Science*, cited in *New Scientist* 1984; p. 57 (13 Sept).
11. Rosenblum M.   1973   Chordata copy editor: a rare species. CBE Newsletter 1973, No. 6: p. 3 (December).
12. Young B.   1975   Manuscript editing: talent, craft and sense of order. Scholarly Publishing 1975; 6:229–233.

13. O'Connor M. 1978 Editing scientific books and journals: an ELSE-Ciba Foundation guide for editors. Pitman, London, 1978 [published in the United States as The scientist as editor. Wiley, New York, 1979].

14. DeVivo A. 1975 Copy editing standards at the American Psychological Association. IEEE Transactions on Professional Communication 1975; PC-18:141–144.

15. Forscher BK. 1985 Preferred background for manuscript editors: English or science? CBE Views 1985; 8(3):5–7.

16. Hall C 1983 Editing for everyone. National Extension College, Cambridge, UK, 1983.

17. Butcher J. 1981 Copy-editing: the Cambridge handbook, 2nd ed. Cambridge University Press, Cambridge, 1981.

18. Judd K. 1982 Copyediting: a practical guide. Kaufmann, Los Altos, CA, 1982.

19. CBE Style Manual Committee. 1983 CBE style manual: a guide for authors, editors, and publishers in the biological sciences, 5th ed. Council of Biology Editors, Bethesda, MD, 1983.

20. Cochran, W. Fenner P, Hill M (eds.). 1979 Geowriting: a guide to writing, editing, and printing in earth science, 3rd ed. American Geological Institute, Alexandria, VA, 1979.

21. Follett W. 1974 Modern American usage. Warner, New York, 1974.

22. Fowler HW. 1965 A dictionary of modern English usage, 2nd ed. revised by Sir Ernest Gowers. Oxford University Press, Oxford, 1965.

23. Strunk W, White EB 1978 The elements of style, 3rd ed. Macmillan, New York, 1978.

24. Hodges JC, Whitten ME. 1984 Harbrace college handbook, 9th ed. Harcourt, Brace Jovanovich, New York, 1984.

25. Nesfield JC. 1963 Manual of English grammar and composition, revised by F.T. Wood. Macmillan, London, 1963.

26. Huth EJ. 1986 Medical style and format: an international manual for authors, editors, and publishers. ISI Press, Philadelphia, in press.

27. BioSciences Information Service. [Annual] Serial sources for the BIOSIS Data Base. BioSciences Information Service, Philadelphia.

28. Index Medicus. List of journals indexed in Index Medicus [issued annually]. National Library of Medicine, Bethesda, MD.

29. Ulrich's International Periodicals Directory. Updated frequently. Bowker, New York.

30. Current Contents. [Several editions, reproducing journal contents pages; published weekly]. Institute for Scientific Information, Philadelphia.

31. Science Citation Index. Institute for Scientific Information, Philadelphia.

Chapter 2

Handling Manuscripts: Practices and Principles

In some respects the scholarly journal is the scientific establishment's ledger of achievement and roll of honour rolled in one. — B. Cronin[1]

This chapter and Chapters 3–8 refer mainly to preparing articles for publication in journals — but the approach is similar if you are working on multi-author books or working directly with authors. If you are a book editor or an authors' editor, don't stop reading yet. But turn to p. 15 if logging manuscripts in when they reach the editorial office is not part of your job.

In addition to logging-in, this chapter on administrative and other aspects of copyediting covers publication procedures you may need to know about, including writing letters to authors and dealing with copyright forms and permissions. Also discussed are instructions to authors; house style and style sheets; nomenclature and terminology; how much copyediting to do and in what order; copy marking; addressing queries to authors; and assembling and mailing manuscripts.

Processing Manuscripts for Journals

Procedures for handling manuscripts for journals vary with the history of the journal, its size (number of pages and income), and whether it is owned by a large scientific or medical society, a commercial publisher, or a small society. At one end of the spectrum a society publishing a large-circulation journal, or several journals, may have a central editorial office with a managing editor who takes care of administration and oversees all aspects of copyediting, typesetting, and printing. Commercial publishers may arrange or pay for similar services for their editors, whether at the publishing house, in the editor's place of work, or by freelances.

At the other end of the spectrum a small scientific society with a single journal may use a freelance copyeditor, but sometimes the scientific editor does the copyediting and nearly everything else too, being "the recipient of all submissions, the publisher's reader, the distributor of papers to editorial board members or to referees, the final arbiter of the acceptability of an article, and the writer of countless letters explaining to authors why their papers need modification or why they are unacceptable."[2] (For more on what editors do see Bishop,[3] Morgan,[4] and O'Connor.[5])

Because procedures in editorial offices vary so much, the rest of this section is descriptive, not prescriptive. If you are responsible for logging-in, follow the system already laid down by your journal, or—if you are the first copyeditor on a new journal—devise a system based on the methods described here.

Logging-in

Whatever the size of the journal, all manuscripts have to be logged in or registered as soon as they arrive. In larger editorial offices clerical staff do this, but on small journals it may be the copyeditor or the editor who keeps track of manuscripts until they are published.

Logging-in, or registration, consists of assigning a reference number to the manuscript or to its electronic equivalent—a compuscript on a floppy disk or magnetic tape, accompanied by a printout. The reference number is placed on a manuscript's title page or on the first page of printout, and sometimes also on the abstract and on all tables, figures, and legends. The date of receipt, the author's name, and the title or a short title are then entered on a chart or control card or in a word processor/computer record (Fig. 5).

After the manuscripts have been logged, it is usual to number the pages if they arrive unnumbered and check whether all the expected parts of the manuscript or printout have been received, such as:

Title page (with title, short title, authors' names and addresses, name and address for correspondence)
Abstract
Key words
Bibliographical identification
Text pages, complete and consecutively numbered
Acknowledgments
Footnotes (non-bibliographical), appendixes
Reference list
Tables, numbered, with titles
Figures, numbered, with legends

Log. no.	Date rec'd	Author(s) name(s)	Short title of MS	Type of contribn	No. of MS p.	Acknowl. sent
87/ 8	3 Jan	Whizzkid et al.	Burning boats	Article	25	4 Jan
87/ 9	3 Jan	Riter & Reeder	To moon & back	Article	15	4 Jan
. . .						

MS to ed.	MS to/from refs.		MS to au.		Revised MS		MS to typeset
	Ref. 1	Ref. 2	To revise	Rejtd.	From au.	To copyed	
4 Jan	Smith 10 Jan 24 Jan	Jones 10 Jan 2 Mar	10 Mar		14 Apr	15 Apr	12 May
4 Jan	Braun 10 Jan 21 Jan	Green 10 Jan 28 Jan	31 Jan		14 Mar	15 Mar	2 Apr
. . .							

Pub. date — estd.	Proofs 1			Proofs 2		Pub. date, vol., issue, p. nos.
	In	Fr. au.	Out	In	Out	
Sept	12 Jun	19 Jun	20 Jun	15 Jul	22 Jul	Sept 87; 112(9): 655–9
July	2 May	9 May	10 May	1 Jun	8 Jun	July 87; 112(7); 489–92
. . .						

Figure 5　Sample columns for a progress chart/computer record for journal contributions.

As a check in case anything goes astray, the number of text pages and other items may be logged in too. The words or characters may be counted, though this is more likely to be done at a later stage (see Chapter 5). If the journal asks for two or more copies of manuscripts, usually for sending to referees (reviewers), the reference number is put on the title page or cover pages before the copies are sent to the referees chosen by the editor.

When logging-in is finished, the editor, or the copyeditor or administrative secretary, acknowledges receipt of the paper and tells the author what the reference number is. As each step towards publication is completed, the date is entered on the chart or other record, which then shows what stage any manuscript has reached and what remains to be done with it. The manual or electronic record can also be used to show the editorial lag (time from receipt of a manuscript until it goes to the typesetter) and the publication lag (the total time from receipt to publication). A separate chart may be kept to show how many pages of a journal issue have been filled and what space remains.

Information about the manuscript may also be written on a control or transmittal sheet that can be filed with each paper and the correspondence related to it (a copy of the title page is sometimes used as the starting point for this sheet). The transmittal sheet travels with the manuscript on its journey into print, instructions for the typesetter being added to it as necessary.

Another set of index cards or word processor/computer records is used for the referees. Details of manuscripts sent to each referee can then be entered, showing the date a manuscript was mailed, the date it was returned, and the action recommended.

Folders or envelopes for each manuscript and the correspondence relating to it are usually kept in numerical order in filing-cabinet sections marked, for example, "Pending," "Accepted," "Rejected," and "Published" (the last two sections are needed because many journals keep manuscripts for a year or longer in case queries arise). Folders are moved from section to section as different stages are completed.

First Steps Towards Publication

After being logged-in, most manuscripts are sent to one or more referees or reviewers who advise the editor on their suitability for publication. Articles accepted for publication are then often returned to the authors for revision in accordance with the referees' or editor's criticisms and suggestions. It is worth noting here that some journals allow or even expect referees to suggest changes in grammar or style. Others ask them not to touch these details; even so, referees often make suggestions that copyeditors later find useful.

Accepted manuscripts may be copyedited either before or after the authors revise the refereed version, depending on journal practice. If the revision requested by the editor and referees is minimal, copyediting is better done before authors revise their work. But if extensive revision is suggested, the copyeditor should work on the revised version — there is no point in copyediting a manuscript that may be changed radically or even submitted to a different journal. (See Fig. 6.)

Manuscripts that are not copyedited until they have been revised to meet editorial criticisms may be returned to the authors a second time, so that authors can check them before the typesetting stage and, if necessary, have them retyped in machine-readable form (in a type face that can be scanned by an optical character recognition [OCR] machine) or in camera-ready form. For both OCR and the camera-ready method the manuscript has to be typed within specified measures, or typed on special paper printed with guiding lines for the typist. For OCR a limited number of type faces are acceptable. The manuscript is electronically scanned and recorded dig-

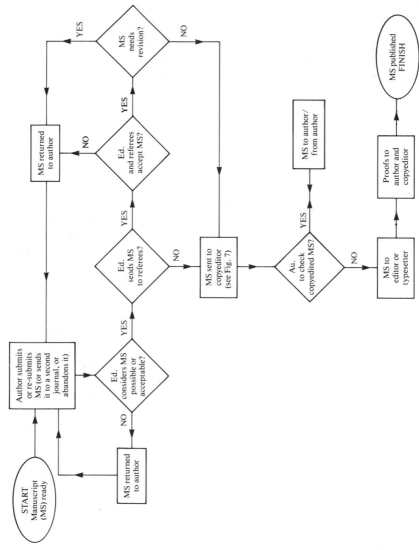

Figure 6 The editing, refereeing, and copyediting stages in the publication pathway.

itally for computerized typesetting. In the camera-ready method the typed pages are photographed to produce the film from which the printing plates are eventually made. If articles intended for printing in these ways are copyedited at all, copyediting has to be done before the author has the final version typed. Alternatively, the copyeditor or someone else in the editorial office may retype corrected sections and paste them in instead of the original lines.

The practice of getting authors to re-check copyedited manuscripts before the typesetting stage is a good one, both ethically and economically, especially when the editor has no time or no opportunity to go over the copyeditor's changes. It is also sometimes used to speed up production by doing away with the need to send proofs to authors. Authors who aren't given a chance to review the copyedited version are more likely to make expensive alterations in proof, some of which will certainly be justifiable. As every surviving copyeditor knows, it is only too easy to change the meaning when trying to make the message clearer.

Letters to Authors

If the copyedited typescript or a copy of it is returned to the author before the typesetting stage, as recommended above, the copyeditor – or the editor, depending on journal practice – usually sends a covering letter to the author, calling attention to the changes and queries. Even if a form letter is used, it should be a tactful one. Authors invest a lot of time and effort in their work and in the manuscripts they write about it; not surprisingly, they may resent anyone else's tampering with their way of writing, especially if the manuscript has already been revised in response to the referees' comments. It is better for copyeditors to regard their changes as suggestions only and to say so in the covering letter than to arouse authors' aggressive or possessive instincts by implying that they write badly: copyediting "corrections" have been known to make matters worse rather than better.

If authors don't see the copyediting changes until proofs reach them, a suitable covering letter or form should be sent with the proofs. The letter can be used to call attention to the editor's/copyeditor's changes and queries, as marked on the copy of the typescript sent with the proofs or – occasionally – as marked on the proofs by the printer.

Copyright Forms and Permissions

When manuscripts are accepted for publication, it may be the copyeditor's job to confirm that authors have either assigned the copyright to the publisher or signed a "license to publish." The publisher can then legally handle requests to reproduce articles, or parts of them, without having to track authors down to get their signatures, perhaps years after publication. In return for agreeing to such an assignment or license, authors are often allowed to reuse their own material elsewhere without having to ask permission each time, provided they include a reference to the original source. A copyright assignment form or license-to-publish form is usually sent to an author with the letter of acceptance, but sometimes forms of this kind are printed in every issue of a journal. The copyeditor may have to ensure that the form has been completed and that all the necessary signatures are included if there is more than one author. These forms can then be filed in the editorial office or sent to the publisher, according to house practice.

The copyeditor may also be responsible for asking authors to obtain written permission ("releases") from other publishers to reproduce their own or anyone else's previously published material, such as tables, illustrations, or substantial portions of text, if releases for these have not been submitted with the manuscript. "Substantial" has not yet been satisfactorily defined in American or British copyright law: a useful rule of thumb for scientific papers is that permission should be obtained for quotations longer than 100 words, or 5% of the total, whichever is less. If the copyright-holder requests a special credit line, the copyeditor has to make sure that this is included in a legend or footnote (see Chapter 6). If no particular credit line is specified, "Reproduced, with permission, from Smith 1985" can be used, with a full reference in the reference list.

For articles submitted to medical journals, copyeditors may have to make sure that authors have obtained written permission from patients to use photographs in which the patients are or might be recognizable.

Copyeditors may also have the related job of handling requests for permission to reproduce material previously published in the publication they work for. People making such requests usually send two copies of a permissions letter for signature, the second being for the recipient's files. If a second copy has not been sent, requests can be dealt with by endorsing the original letter with a specially designed permissions stamp, photocopying the letter for the files, and returning the original to the sender. Requests to use a whole article or a large part of it should be referred to the editor or publisher. A fee may be charged if the borrowers are likely to make a profit from an anthology of previously published material or if they intend to distribute copies of an article to promote the sale of drugs or other products.

If the journal publisher is registered with the Copyright Clearance Center in the United States, or a similar center elsewhere, agreed fees for photocopies of articles are collected by the Center or other organization. A code is printed in the journal's masthead or on the first page of each article to allow the CCC system to be administered. The copyeditor may be responsible for adding this code to the manuscript before it is typeset. (See Chapter 7.)

More Preliminaries to Copyediting

So much for the background procedures. Before you begin the main work of copyediting, study your journal's instructions to authors, its house style, and the recognized terminology of the disciplines covered.

If you are an authors' editor, your first job may be to help authors to decide which journals should be their main targets. Is the manuscript of general or specialized interest? Which journals have healthy circulations and good reputations for fair, helpful, and prompt refereeing, as well as for publishing articles of high quality? Which journals publish articles quickly after accepting them? Which have handling charges in addition to page charges? Which reproduce illustrations well? And which have their contents abstracted in the principal databases or have article titles listed in *Current Contents*[6]? Look carefully at recent issues of the journals you think are most appropriate, read their aims and scope (usually at the beginning of the instructions to authors), and assess the standing of the editor(s) and editorial board members to the best of your ability before you make your recommendations to the author.

Instructions to Authors

Instructions to authors usually appear in every issue of a journal or in the first issue of the year. They tell authors, in greater or lesser detail, how to prepare manuscripts for that particular journal. Sometimes they give advice on terminology, the treatment of units, the layout of tables, and so on, and they nearly always show how references should be handled in the text and in the reference list.

The instructions will be particularly detailed if the journal is prepared from manuscripts typed in an OCR type face or from camera-ready copy provided by the author. If the journal uses authors' disks or computer tapes (compuscripts or soft copy) for typesetting, there will be special instructions on preparing these too. Authors are usually asked to provide printouts as well as disks or tapes, and to use a letter-quality printer to prepare the copy. Unnumbered and unseparated fanfold paper, complete with sprocket-hole strips, will probably be forbidden but may nevertheless appear on editors' and copyeditors' desks.

House Style, Style Sheets, Progress Charts

House style, built from a mixture of experience and editorial whims, is the journal's collection of decisions on preferred and prohibited terms; on what is to be abbreviated, capitalized, italicized, or hyphenated; and on which spelling or punctuation is to be used when equally acceptable alternatives exist. Its purpose is to keep journal style reasonably consistent, no matter how many copyeditors there are. If house style is kept to a practical size, it saves the copyeditor's time and provides a valuable last line of defense when authors complain that their favorite spelling or punctuation has been changed. The Second Edict of Copyediting, of course, is:

> ### FIND GOOD AUTHORITY
> ### FOR EVERY CHANGE

In addition to consulting the established house style of your journal, build up a style sheet of your own by ruling a page or pages of typing paper into boxes for the different letters of the alphabet and for items such as tables, numbers, and abbreviations (see Ref. 7, p. 60–62, or Ref. 8, p. 44–55). List the decisions you make and note where the word or phrase first appears, and in which manuscript, in case you change your mind later on. Some of your decisions on style may find their way into the more formal collection of house style.

To keep track of your work on manuscripts, a chart similar to the logging-in record is useful, with columns for the various steps in copyediting.

Nomenclature and Terminology

Most editorial offices collect published lists of recommended nomenclature and terminology in the subjects they deal with. These recommendations may not be easy reading, especially if the subject is not your own, but the effort will be worth it in the long run and will give you an insight into the discipline you are dealing with. Nomenclature is revised as research progresses and you'll need to watch for announcements of new versions. If no lists of recommended nomenclature are available in the office, consult the *CBE Style Manual*[9] (p. 155–240) or a manual in the appropriate discipline.

How Much Copyediting, in What Order?

The depth of editing you are expected to do should be specified by your employers — but you may be left to develop your own approach. Checking for completeness (C) and marking up the format (F) might be the minimum level required (see Fig. 2, part 1, and Fig. 4, part 3, in Chapter 1). Marking mechanical style (M) might be added to these two processes at the next level of editing (Fig. 4, part 2). Correcting the language (L) could be added at the third level, and dealing with substantive editing (S) at the fourth. That is, level 1 consists of $C+F$, level 2 of $C+F+M$, level 3 of $C+F+M+L$, and level 4 of $C+F+M+L+S$ (see Ref. 10 for other possible levels of editing).

If you are an authors' editor, you can either talk to authors about the levels of editing they want you to do or find out by trial and error what each person needs.

If you are a freelance copyeditor, ask your clients for some general guidelines on how much work they expect you to do. If necessary, explain what the different levels of editing include and how your charges for them vary. If you are asked to do level 1 editing for a manuscript that turns out to need level 2 or level 3 editing before it is fit for typesetting, explain this to the client as tactfully as possible. Where appropriate, point out that the cost of typesetting and proof correction will rise out of all proportion to the amount saved by restricting copyediting to a lower level.

As a minimum, any manuscript passed for typesetting or printing should be grammatically correct, sensibly punctuated, properly spelled, and clearly marked up to make it conform to the journal's usual format and style.

If you are responsible for the wider and wilder shores of substantive copyediting, do this more fundamental work on a manuscript either before technical editing or at the same time (see Chapter 3). Your work on the details of grammar and mechanical style could be wasted if substantive editing is done after technical editing.

Whether you are doing technical editing alone or technical plus substantive editing, one way to proceed is to tackle everything at once in two readings of each manuscript, perhaps followed or preceded by a separate stage for cross-checking and styling the references. Another way is to edit the language first (at the same time as substantive editing, if you are responsible for this), and then deal with format and mechanical style, again with two readings and possibly a separate stage for the references (see Fig. 7 for the possible combinations). During your first working-over of the manuscript, deal with all the items listed in Checklists 1 and 2 (p. 40 and 61). Use the second reading for polishing up the details and checking your changes— failure to delete all the words one means to delete is common ("a limited number of type faces are be acceptable" appeared in the draft manuscript of this book a few pages ago); failure to go back and write in revisions where you have whited words out is common too. Keep track of what you've done by checking off items such as first reading, second reading, and reference cross-checking on your progress chart.

A problem with difficult subject-matter in science is that even when manuscripts need very little copyediting, it still takes a long time to work through them and confirm that they really are in good order. Some employers, however, expect copyeditors to deal with a certain average number of manuscript pages every working hour. Five to ten double-spaced pages an hour is a typical rate if level 2 editing as defined above is all that is re-

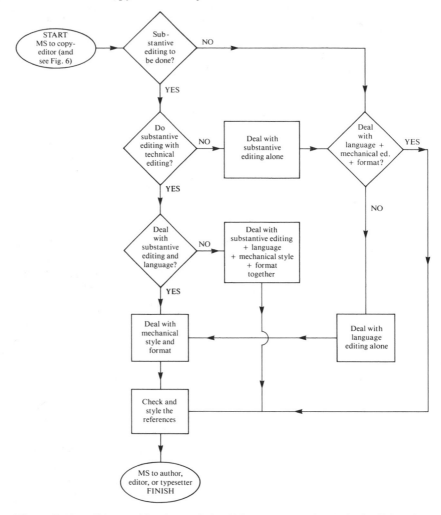

Figure 7   Possible combinations of the different types of technical editing (language, mechanical style, format) with (or without) substantive editing, with reference checking and styling as a separate step in mechanical style.

quired. One or two pages an hour is more realistic for difficult text if language editing and substantive editing are specified, especially if the manuscript is in a language that is not the author's mother tongue.

However much or little time you can spend on each manuscript, be sure to keep to any deadlines that have been set and observe the Third Edict of Copyediting, which is the corollary of the First (p. 7):

> ### MAKE ESSENTIAL CHANGES ONLY

To this end, ask yourself these questions:

1. Is a change really necessary here?
2. Why is it necessary?
3. Is my version an improvement — or a backward step?

New copyeditors — or all of us — should write these questions out in large letters and pin the paper where they can see it at all hours of the working day. They (we) might also note the Fourth Edict:

> ### PROTECT READERS FROM AUTHORS
> ### AND AUTHORS FROM THEMSELVES

That is, make sure that the manuscript is as readable and correct as possible, and that it says what the author wants it to say, insofar as you can decide what that is.

Marking Manuscripts

There are two kinds of manuscript-marking in copyediting. *Mark-up*, dealt with in Chapter 7, consists of marking the manuscript or compuscript to show which type faces, sizes, and weights (roman, italic, or bold) should be used, how headings should be spaced, how much space the type should occupy on the printed page, and so on, in accordance with specifications originally drawn up by a designer. *Copy-marking*, dealt with in this section, consists of changes to the language or mechanical style of the author's original text, these changes being made in or between the lines of text. Keys or codes showing the rank of headings are also usually inserted at the copy-marking stage, although keying belongs, strictly speaking, to mark-up. All your markings must be crystal clear to the typesetter and to the author, or to the person (perhaps you) who has to enter corrections on-screen. If your usual handwriting is less than ideal, try to develop an easy-to-read script for correcting and marking manuscripts and proofs and for writing queries to authors. Always write on manuscripts in the same direction as the text is typed and on the same side of the page, never at right angles to the typed lines or on the back of the page; both these habits make life hard for the typist or whoever keyboards the changes. For the same reason, avoid making changes close to the bottom of the page. If there isn't enough room on the page for long inserts, type them on a separate page clearly labeled "insert A for page 9, line 17" and write "insert A attached" on page 9. And if the manuscript or printout is likely to be photocopied later, don't write too near the edge of the page.

Editing on a screen (on soft copy) is slower, more liable to error, and much more tiring to the eyes than using a pen or pencil on paper. Instead, edit the paper copy (hard copy) and either make the changes on the screen yourself or give them to a keyboard operator to make — or it may be journal practice to send the edited manuscript back for the author to make the changes.

Copy-marking

The marks for deleting or inserting material in manuscripts (Table 1) resemble standard proofreading marks, with one major difference in the way they are used. On proofs, all corrections and other changes must be made in the margins as well in the text (see Chapter 9). In manuscripts, make corrections and changes *in or between the typed lines*, where the typist or typesetter can see and follow them easily; do NOT make corrections in the margins unless there is no space left between the lines. Treat a double-spaced or treble-spaced printout from a compuscript in the same way as an ordinary manuscript, but put a cross or other mark in the left margin beside each line containing a change or instruction. If the printout is single-spaced, leaving no room for corrections between the lines, make changes and corrections in the margins and use lines or loops and arrows to show the keyboarder where these changes should be made in the text (this is easier for many keyboarders than trying to follow a sequence of corrections and marks in the margins). On your first reading of the manuscript, put a pencil mark in the margin against lines where you cannot decide whether a change is needed; come back to these points after you have read the whole paper.

Make changes or give instructions on manuscripts consistently. If you correct a word with an incorrect initial capital the first, second, and fifth times it appears, but not the third, fourth, and sixth times, the typesetter can't be expected to guess which version is intended. With a compuscript, however, you might more safely give one marginal instruction at the first appearance of an incorrectly styled or misspelled word: "change Nature to nature all through." For Greek letters and well-known but little-used signs or symbols (prime signs, for example) it is also usually sufficient to write the name of the letter or sign at the first appearance only, at least in short manuscripts submitted to journals: good typesetters recognize these as easily as they recognize any other characters. Where the letters "l" and "O" could be confused with the numerals 1 and 0, distinguish between these at each

Table 1 Marks used in or between the lines of text in correcting manuscripts[a]

Instruction	American	British	Comments
Leave unchanged (stet)	· · · · ·	- - · - -·	Place below characters or words incorrectly deleted
Insert	∧	⋏	
Delete	⤸ or / or ▬	⸦ or ⊢⊣	Through characters or words to be deleted
Delete and close up	⤸̂	⤸̃	
Set in italics	▬▬▬	▬▬▬	Place below characters or words
Set in capital letters	≡≡	≡≡	As above
Set in small capitals	≡	≡	As above
Set in bold type	∿∿∿	∿∿∿	As above
Change capitals to lower-case letters	A̸	A̸	
Superscript (superior) character (above line)	∨a∨	∨a∨	
Subscript (inferior) character (below line)	∧b∧	∧b∧	
Start new paragraph	⏀	⌐	
Run on (no new paragraph)	⟿	⟿	
Transpose characters or words	∽	∽	
Center material	⌐ ⌐	⌐ ⌐	
Indent	⊡, ⊡⊡ or ⬛, etc.	⌐	Indicate size of indent (1 em, 3 mm) as needed
Cancel indent	⌐	⊢⌐	
Move to right	⊐	⌐⊢→	Use in tables
Move to left	⊏	⊢⊏ ⊐	Use in tables
Close up (delete space between characters)	⌣	⌣	
Insert space between characters or lines	⎸, >	⎸, ⟩⊣	Give size of space if necessary

[a] These marks are for use in manuscript text—in the lines of text, not in the margins. The first mark is a general one; the next 10 marks indicate deletion, insertion, or substitution; and the remaining marks show positioning and spacing. For a full set of marks for use on proofs see Chapter 9.

appearance, if necessary, or write a general instruction: "Al = cap. A & l.c. el all through" (Al being the symbol for aluminum, in this example). Long manuscripts, however, may be set by several keyboarders; in these it is better to repeat instructions wherever they are needed.

Your journal may ask you to make all marks and corrections on manuscripts in ink for the typesetter, but to begin with you may prefer to use a soft black pencil or a pen with erasable ink that allows you to revise your changes on your second reading. Pens with erasable ink, however, may have blunt ballpoints that are not ideal when you have to write very clearly between typewritten lines. If you use a pencil, you may later have to spend time inking in your changes to make them acceptable to the typesetter; this is not usually regarded as a cost-effective procedure. Ink corrections are perhaps most easily changed by using correction tape that exactly fits between double-spaced lines. Test the various possibilities and see which method suits you best. If you use blue ink or a blue pencil, make sure that the marks are reproduced clearly on photocopies — blue sometimes comes out faintly or not at all (the same applies to some reds).

Writing Queries for Authors

Write your queries for authors clearly in the margins of the manuscript (but see next paragraph). Circle the queries, which should be flagged "Author," or "Au." Don't write too near the edge of the page or some words may be chopped off if the manuscript is photocopied.

The margins may have to be kept clear for instructions to the typesetter, or they may be too narrow for the queries. If so, use Post-it (peel-off) notes or flags. Fix the gummed part of the flag to the back of the page and fold the flag forward over the margin before writing your question. Write the page and line number on each flag in case it peels off prematurely. The drawback to using flags is that authors may remove them, leaving you with no record of your queries if the flagged manuscript cannot be photocopied before you send it to the author.

A third method of addressing questions to authors is to list queries on a separate sheet of paper, linked to letters or numbers in the margins. The list can be sent to the author either when the paper is ready for typesetting or with the proofs. If the author doesn't see proofs, ask for a reply to your queries well before the proofs are due, so that you can make the changes when they arrive from the typesetter.

Keep your queries very short. Mini-essays on the finer points of grammar or the ten other possible ways of writing a sentence are unlikely to be appreciated. Word your queries concisely and in such a way that authors give unambiguous replies and are neither aggravated by the tone of your questions nor aggrieved or amazed by your ignorance. Authors quickly lose faith if you display your lack of what they regard as elementary knowledge of their subject, so try to disguise shortcomings of this kind. Write "Change OK?" rather than "Original wording pompous. Is this what you mean?" Do not ask "Do you mean X or Y?"—to which you may get the reply "Yes." Do not write, beside a reference in the bibliography, "Not in text" (or NIT); "Where in text?" is preferable (see also Chapter 6). Say "Please reword" or "Suggest clarify," not "I don't understand why 2 + 2 = 4"—but wherever possible it is a good idea to write your suggested revision on the manuscript for the author to approve or amend, rather than ask for passages to be rewritten without showing how this might be done.

In your contacts with authors, you must not be intimidated by Nobel Prize winners, members of national academies, Fellows of the Royal Society, and other stars in the scientific firmament. But do not be aggressively sure that your version of authors' prose is better than their own: they have already forgotten more about their subject than you are ever likely to know, and it is they who, in the end, have to stand by the words that appear in print under their names. A Fifth Edict of Copyediting may be useful:

> ARGUE SO FAR AND NO FURTHER;
> THEN GIVE WAY GRACEFULLY

A Sixth may be helpful too:

> RESPECT AUTHORS' FEELINGS—
> PRAISE BEFORE YOU CRITICIZE

While the Seventh is essential:

> ALWAYS COMMENT CONSTRUCTIVELY

Use your common sense and read the manuscript carefully before asking a question you ought to be able to answer for yourself. If the author has just mentioned experiments featuring the chorioallantoic membrane, for example, or observations made on mud deposits beside the Ganges, don't ask in what species or country the work was done.

Cleaning Up Manuscripts

If authors are sent their copyedited manuscripts to check over before the typesetting stage, make sure—when (or if) the manuscripts come back to you—that all your queries have been answered. Then clean up the margins for the typesetter, as just described. If any questions remain, mark these in colored ink in the margins, address them to "Au" or "Author," and circle them. If there is no space to do this, or if the margins must be left completely clear, make a list of the queries and send this to the author either immediately or with the proofs, as mentioned above.

Assembling Manuscripts for Typesetting and Mailing

When you have done your best, or worst, with manuscripts, (re)assemble them for typesetting in the order in which most typesetters prefer to receive manuscript copy. That is, put the title page first, then the text, references, tables, figure legends, and figures, and recheck that the pages are correctly numbered. If manuscripts with illustrations have to be sent by mail, cover the figures with thin cardboard and send the whole package out in a strong envelope, firmly taped up, and with the mailing category (Airmail, Foreign Airmail, First-class, etc.) clearly marked.

References

1. Cronin B.   1984   The citation process: the role and significance of citations in scientific communication. Taylor Graham, London, 1984.
2. Sheffield F.   1985   Editing a small circulation journal. Earth & Life Science Editing 1985; No. 25: p. 9.
3. Bishop CT.   1984   How to edit a scientific journal. ISI Press, Philadelphia.
4. Morgan P.   1986   An insider's guide for medical authors and editors. ISI Press, Philadelphia.
5. O'Connor M.   1978   Editing scientific books and journals: an ELSE-Ciba Foundation guide for editors. Pitman, London, 1978 [published in the United States as The scientist as editor. Wiley, New York, 1979].
6. Current Contents. [Several editions, reproducing journal contents pages; published weekly]. Institute for Scientific Information, Philadelphia.
7. University of Chicago Press.   1982   The Chicago manual of style, 13th ed. University of Chicago Press, Chicago, 1982.
8. Judd K.   1982   Copyediting: a practical guide. Kaufman, Los Altos, CA, 1982.
9. CBE Style Manual Committee.   1983   CBE style manual: a guide for authors, editors, and publishers in the biological sciences, 5th ed. Council of Biology Editors, Bethesda, MD, 1983.
10. Van Buren R, Buehler MF.   1980   The levels of edit, 2nd ed. Jet Propulsion Laboratory, California Institute of Technology, Pasadena, CA (JPL Publication 80-1), 1980.

Chapter 3

Substantive Editing

Substantive editing—"rewriting, reorganizing, or suggesting other ways to present material"[1] (p. 51)—is usually a regular part of the job for authors' editors. Other copyeditors may be asked not to do any substantive editing at all (Chapter 1), possibly because employers think edited and refereed manuscripts need nothing more than technical editing, or because they find it safer or cheaper to specify technical editing alone. Substantive editing certainly requires an understanding of what authors are saying, and preferably plenty of copyediting experience, but this kind of editing overlaps with technical editing, especially language editing, as well as with creative editing (Fig. 1, p. 2). Many copyeditors therefore find themselves doing substantive editing even when it is not strictly part of their job.

This chapter outlines the main features of manuscripts that are usually dealt with under the heading of substantive editing: manuscript titles, abstracts, key words, headlines and headings, logic and order, redundancy, references, tables, and figures (see Fig. 3, p. 3). If you are responsible for this kind of editing, work on these features either before or at the same time as technical editing (see Fig. 7, p. 22). Ideally, you should read the manuscript through to get a general idea of its message before you tackle the items described here.

Titles of Manuscripts

If your journal has a limit of 12 or so words for titles, has the author kept to this length? If not, can you remove words or phrases such as "Notes on" or "A study of" without sacrificing the sense? Can shorter or more concise expressions be used? Can you suggest other ways of making a title easy to understand for readers leafing through journals, and easy to retrieve for those consulting bibliographic databases? The example below is unlikely to attract a journal browser, though it may be easily retrievable:

> Human monoclonal anti-keyhole limpet hemocyanin antibody-secreting hybridoma produced from peripheral blood B lymphocytes of a keyhole limpet hemocyanin-immune individual

If a monster of this kind jumps out of the page at you, read the beginning of the paper before you try to tame the title by making it concise, accurate, and informative. My editorial colleague at the Ciba Foundation, Julie Whelan, who trapped the real-life horror above, suggested the following friendlier version after reading the introduction to the article:

> B lymphocytes from a man immunized with keyhole limpet hemocyanin (KLH) produce a hybridoma secreting a monoclonal anti-KLH antibody

This version illustrates the rule that abbreviations should not be used in titles without being explained, unless they are abbreviations that have taken over from the original terms (DNA, A.D.). Note, however, that some journals do not accept titles of this kind — statements in the present tense. Another possibility, provided that "keyhole limpet hemocyanin" is not an essential key term for indexing the paper, would be:

> Production of a human hybridoma secreting a monoclonal antibody of predetermined specificity

Informative titles telling the reader what is in a paper are usually preferable to *indicative* titles describing what the paper covers. Compare the following:

> Activation of ion channels in locust muscle by amino acids [informative title]
>
> Membrane permeability in insects [indicative title]

If a title seems too general, read the abstract or summary and the introduction; then try to make the wording of the title more specific. Put the most important terms first, avoid using strings of modifiers and too many prepositional phrases, and add subtitles if necessary and if journal style allows:

> Ion channels in locust muscle: activation by amino acids

If you make anything other than the most trivial changes to titles, check your version with the author or editor before the manuscript is typeset.

Abstracts

An abstract is defined as an abbreviated, accurate representation of a document, without added interpretation or criticism, preferably placed at the beginning of an article or chapter.[2] A summary is a restatement of the main findings and conclusions, but not necessarily of the purpose and methods described in a document, and it is usually placed at the end "to complete the orientation of a reader who has studied the preceding text."[2] Some journals, however, call their abstracts "summaries."

Like titles, abstracts for journal articles other than review or other long articles are more useful if they are informative rather than indicative (descriptive). An informative abstract should tell readers the purpose of the work, indicate the methods used, and summarize the results and conclusions, all within a brief space and without repeating or paraphrasing the title. That is, an abstract should answer the questions "why did you start, what did you do, what answer did you get, and what does it mean anyway?" (as Bradford Hill put it[3]) — while remaining as concise, accurate, and informative as possible. Compare this informative abstract:

> The usefulness of the work done by authors' editors at this institute was questioned in 1984. We therefore compared the effectiveness of 100 manuscripts submitted for publication after an authors' editor had worked on them with the effectiveness of 100 control (untreated) manuscripts. Effectiveness was based on the number of substantive changes made by copyeditors of the target publications. Treated manuscripts averaged one change per page compared with 12 changes per page for untreated manuscripts. Proofreading time and the time from submission to publication were also reduced for the treated manuscripts. An extra authors' editor has now been employed and the salaries of this category of staff have been raised.

and this indicative version:

> As the effectiveness of the work of authors' editors was not known, it was decided to investigate this question. A survey of manuscripts handled or not handled by authors' editors was undertaken. This survey is described here and the results are discussed. It is concluded that the employment of authors' editors is of benefit.

If the informative kind of abstract is preferred by your journal, can you do anything to convert a descriptive abstract like the second example into an informative one like the first example? If not, consult the editor, or ask the author to provide an informative version. A hybrid type called an "informative-indicative abstract" may be acceptable for long papers, especially review papers.

Check, too, whether your journal's requirements for length and format (a single paragraph? numbered paragraphs or sentences?) have been observed. Is use of the first person permitted? Have abbreviations, acronyms, and unfamiliar terms or symbols been explained at first mention? Has the abstract been written in complete sentences rather than telegraphese?

Footnotes, tables, diagrams, equations, structural formulas, and references to these and to published work should be banished from abstracts unless there is some special reason for including them. If references to other work are essential, include a brief version of the bibliographical information in parentheses after the name of the cited author ("J Imag Chem 1985; 99:20–29"), and make sure that the reference is cited in the main text too.

Check that all statements in the abstract are dealt with in the article, and that the main findings and conclusions in the article are covered in the abstract. In other words, does the article deliver the promised goods, and does the abstract represent those goods accurately? (See Cremmins[4] for more on abstracts.)

Key Words

A short list of words referring to the main points covered in a manuscript is sometimes printed after the abstract to help readers who are searching for articles on particular topics, or to help whoever constructs the index (Chapter 8). The words are often chosen from a thesaurus—a list of terms approved by the journal or accepted in the discipline it covers. Has the author supplied these key words if so required? If not, list the terms you think are suitable and ask the editor or author, or both, to approve them, if necessary.

Running Headlines, Headings, and Subheadings

If the journal prints headlines ("running heads") or footlines for each paper, has the author supplied a short title with no more than the stated maximum number of characters? If not, invent one based on the main title or on information in the abstract or introduction. Ask the editor or author to confirm that your version is acceptable, if you have any doubts about it.

Section headings and subheadings should be suitably specific; "Size" and "Epidemiology" on their own are not informative.

If section headings are lacking in places where you think they would help readers, insert your own suggestions and ask the author to approve them or amend them.

Headings for a series of sections on related topics should match each other in style. "Silicon–water chemistry," for example, should be followed by "Phosphorus–water chemistry" in preference to "The chemistry of phosphorus–water." Make headings parallel in this way when necessary.

Do not assume that readers will read headings. The text after the heading "Silicon–water chemistry" should read "Silicon–water chemistry can be considered from several angles," not "This can be considered from several angles." If sentences immediately after headings begin with "This," replace "This," if necessary, with wording similar to but preferably not identical with the wording of the heading.

Logic, Order, and Much Else

As you read a manuscript, ask yourself whether the contents of sections match their headings. Results often creep into the Methods section, for example, or discussion may appear in a section headed Results (if Results and Discussion are separate sections). If you think any of the text should be moved, or that sections such as Results and Discussion should be amalgamated, consult the editor or ask the author whether these changes are appropriate.

The various sections of a manuscript should not only correspond to their headings but also be logical internally, logically ordered in relation to each other, complete in themselves, correct, and emphasized in proportion to their significance in the whole article. Authors know their own work so well that they (and referees doing similar work) often don't realize when a step in the argument has been left out. Look at the argument critically as you read. If you find gaps, other non-specialist readers will probably be puzzled too. So make sure that all the steps are there – and that all the necessary facts are included, with the necessary references.

Make sure, too, that statements in different sections are consistent and agree with each other, and that references in the text to material in tables and figures agree with what the tables and figures actually show. Resolve any discrepancies or query them with the author.

Copyeditors are not expected to correct scientific statements – but you can at least look critically at more general comments and check their correctness when possible. Your general knowledge and common sense can be put to good use in rescuing authors from various major or minor *faux pas*, such as attributing quotations wrongly. General MacArthur was not the originator of the line "Old soldiers never die," as an author of my acquaintance apparently thought when he compared macrophages to old soldiers (British soldiers in World War I beat the general to it).

You should also look out for ambiguous terms, for example those with different meanings in different countries. "Billion" is one such word: change it, if possible, so that readers know whether it means a thousand million (10^9), as it does in the United States and France, or a million million (10^{12}), as it usually but not always does in the United Kingdom.

In methods sections, in particular, make sure that complete information is included on instrumental methods. In nuclear magnetic resonance studies, for example, when chemical shift data (δ) are given, the reference compound must be named.

Repetition, Redundancy, Irrelevancy, and Emphasis

Besides making sure that everything necessary is included in a manuscript, remove everything that is not. Look out for repetitious, redundant, irrelevant, or inappropriate phrases or sentences. (In journal articles most repetition is redundant, but not everything that is redundant is repetitious.) Long portions of text that contribute nothing to the argument will usually have been removed by the editor and referees, but if you think any redundant paragraphs remain, bring them to the editor's attention or ask the author, tactfully, if these parts can usefully be removed. In particular, lengthy introductions to papers may need to be cut.

Sections of the Discussion that simply repeat numerical results presented in the tables should be removed, or the Discussion should perhaps be amalgamated with the Results. Conclusions that meander on without getting anywhere may need to be straightened out by removal of sentences of this kind:

> Further investigation of some of the rather meager evidence that exists regarding these issues may help to focus attention on those questions that are in special need of discussion.

Length of Words, Sentences, and Paragraphs

If an author always uses long words or phrases when equivalent short ones exist, change the long terms. Alter "approximately" to "about," "are of the same opinion" to "agree," "it is often the case that" to "often," and so on (see Ref. 5, p. 93–98; Ref. 6, p. 36–37).

Change passive constructions to active where appropriate, and simplify and strengthen the language by liberating the active verbs trapped in abstract nouns (see Chapter 4, sections 3d and 3f). These operations usually make sentences shorter.

Sentences that are long but easy to understand are perfectly acceptable, but if an author uses a succession of very long and difficult sentences, try to break them up for greater readability. If readers are to stay awake, sentences in scientific writing should vary in length between about 8 and 40 words.

The same principles apply to paragraphs as to sentences. Readers find solid blocks of text tiring, so try to break up paragraphs that spread themselves over more than half a typewritten page. On the other hand, paragraphs should deal with one topic, usually introduced in the first sentence, so don't break up the text in unsuitable places — a paragraph, as Fowler[7] points out, is a unit of thought, not a unit of length. And don't make too many paragraphs too short; a series of one-sentence paragraphs may be acceptable in a popular newspaper but is unsuitable in scientific writing.

Keep the readers in mind when you edit the language in this way. The readers of specialist journals are likely to be specialists too, who won't need or want the language to be oversimplified. But readers of interdisciplinary or more general journals will need to have difficult sentences unraveled and unusual or recently invented words explained.

References (see also Chapter 6)

Under-referencing or over-referencing will usually be dealt with by the editor and referees. But if you find that a statement such as "It has recently been reported that X is responsible for Y" has no supporting reference, ask the author to reveal who published this report and to provide the bibliographical details. On the other hand, if authors give more than three references in support of every statement, check whether it is journal practice to include so many, or consult the editor.

Tables (see also Chapter 6)

A table is a way of displaying information, numerical or otherwise, in columns and rows. Check that all tables really are tables and are not better treated as figures, or perhaps as lists within the text — and note that informal (unnumbered) tables are to be avoided because they can lead to make-up problems at the production stage.

Decide, too, whether tables are effectively and economically designed. Tables are expensive to typeset and should be used only when the information presented is extensive enough (at least three rows) to justify their use. On the other hand, tables should not be too long (one typed page, double-spaced, is a reasonable size). If tables have very few entries or if all results are negative, would it be more sensible to put the information in the text and delete the table? A column of zeros, for example, can usually be adequately replaced by a statement that all results were negative. Alternatively, would results included in the text be more effectively presented in a table or tables?

Check whether columns and rows appear in the order that best emphasizes what each table aims to show. Columns whose contents are compared should be beside each other, and control or normal values should usually occupy the first column or row. Would any tables be easier to read (and cheaper to typeset) if they were turned through 90 degrees?

If a table is mentioned twice or more in the text, decide whether it should be placed nearer the first mention (the usual choice) or the second mention, or anywhere else — always provided that the tables and their numbering run consecutively, in the order of the first mention in the text. Write "Table 1" etc. in the margin at the appropriate place and circle this message.

Make sure that table titles represent table contents satisfactorily, that no unnecessary words ("Results from a study of") are used, and that tables in a series have titles matching each other in construction. Likewise, where different parts of a table or different tables in a series give similar information, make sure that the tables are presented in a similar (parallel) form.

In column subheadings, a factor such as "$\times 10^3$" is ambiguous; it may indicate that numbers in the column are to be multiplied by that factor, or that the numbers in the column have already been multiplied by 10^3. That is, the number "5.0" could mean 5000 g if $\times 10^3$ means that numbers in the column are to be multiplied by 10^3, or 0.005 g if $\times 10^3$ means the numbers have already been multiplied by that amount. Remove this source of confusion by changing the heading to a unit of measure — "kg" or "mg" in this example — if you can guess the most likely unit (and ask the author to check it). If clues in the table or text don't tell you what the factor means, ask the author.

If the author has used five or more digits for numerical results, consider changing the unit of measure in the heading to one that reduces the number of digits: for example, if 298 765 (or 298,765) is entered in a column or row referring to millimeters (mm), round this off to the nearest significant figure (298.8) and change "mm" to "m."

In statistical analyses, check that the mean (x) is supported by statistics such as the number of observations (n), the standard deviation (SD), and, where appropriate, the standard error of the mean (SE), and that the last two are identified as SD or SE (see Ref. 6, p. 151: "Standard deviation refers to the variability in a sample or a population. Standard error . . . is the estimated sampling error of a statistic such as the sample mean.").

Probability values (P or p) are often quoted with no indication of the statistical test used to obtain them (for example the χ, t, or F test). If you notice an omission of this kind, consult the editor. Make sure that the signs for "greater than" ($>$) and "less than" have been correctly used with P values; they are often written in the wrong way ($<$ is used when a result is significant).

Figures and Their Legends (see also Chapter 6)

Make sure that figures and their legends, like tables, agree with what is said about them in the text, that they really are figures, not tables, and that the results shown in graphs are not duplicated in tables or the text. Mark the required position of figures in the published article with a marginal note.

Examine photomicrographs (light micrographs, electron micrographs, and so on) and other photographs for their suitability for reproduction. If they are not sharply focused, or if there is poor contrast between the light and dark parts, ask the author whether better copies are available. If at first you are not sure how to judge the quality of photographs, ask the production editor or printer, if you can, to comment on a batch of photographs for you. You will soon be able to recognize which are acceptable and which are not.

Parts of photographs may be redundant in the printed article. The author may have made pencil marks (crop marks) on the edges of the photograph or provided an overlay showing the area that needs to be reproduced. If the author hasn't done this, examine the photographs in relation to their legends and the text, and mark any parts that are superfluous (use crop marks on the edges or back of the photograph, or outline the required area on an overlay). If necessary, ask the editor or the author, or both, whether your suggestions are acceptable. Do the same for line drawings that contain too much detail.

On maps, make sure that place names will remain legible if the maps are to be reduced for publication. If maps are to be redrawn, supply a typed list of the names and separate the different categories (towns, rivers, and so on) if they are to be printed in different ways (all capitals or small capitals, etc.). Check that the names on the map are spelled the same way as any mention of them in the text. If names of rivers or other features are to be printed out of the horizontal, indicate that they should all face the same way, or make sure that they already do so on the map submitted by the author. If north is anywhere other than at the top of a map, insert a north point if it isn't there already. Add scales to maps if these are missing, and make sure that different colors or different kinds of shading are explained by keys on the maps or in the legends, as appropriate (see Ref. 8, p. 51–52).

In histograms, if bars are hatched, make sure that the hatching will be reproduced clearly enough; the hatching should not be too fine.

In graphs, is it clear what factors such as "$\times 10^3$" mean (see "Tables" above) and whether "0" is the point of origin or zero on a numerical scale? (Mark a point of origin as an italic capital O: Ref. 8, p. 53.) If the points on an axis are on a descending scale, mark a break in the axis between the zero and the first point (otherwise the values will jump from 0 to, say, 10^6, 10^5, . . .).

In Conclusion

Substantive editing is more of an art than a craft. The talent for substantive editing develops with experience, like the ability to edit creatively[9] (p. 2). Plenty of work of the kind described in the next four chapters will help this talent to grow . . .

References

1. University of Chicago Press.   1982   The Chicago manual of style, 13th ed. University of Chicago Press, Chicago, 1982.
2. ANSI Z39.14-1971. American national standard for writing abstracts. American National Standards Institute, New York, 1971.
3. Hill A Bradford.   1965   The reasons for writing. British Medical Journal 1965; 2:870.
4. Cremmins ET.   1982   The art of abstracting. ISI Press, Philadelphia, 1982.
5. O'Connor M, Woodford FP.   1975   Writing scientific papers in English: an ELSE-Ciba Foundation guide for authors. Excerpta Medica, Amsterdam, 1975/Pitman, London, 1978 [2nd edition, Wiley, Chichester and New York, 1987].
6. CBE Style Manual Committee.   1983   CBE style manual: a guide for authors, editors, and publishers in the biological sciences, 5th ed. Council of Biology Editors, Bethesda, MD, 1983.
7. Fowler HW.   1965   A dictionary of modern English usage, 2nd ed. revised by Sir Ernest Gowers. Oxford University Press, Oxford, 1965.
8. Butcher J.   1981   Copy-editing: the Cambridge handbook, 2nd ed. Cambridge University Press, Cambridge, 1981.
9. O'Connor M.   1978   Editing scientific books and journals: an ELSE–Ciba Foundation guide for editors. Pitman, London, 1978 [published in the United States as The scientist as editor. Wiley, New York, 1979].

Chapter 4

Language Editing

Language copyediting, like substantive editing, is obviously a vast subject. Here I shall suggest a few principles you can follow, and then point to the main problems you are likely to meet in dealing with spelling, punctuation, grammar, and usage. For help when the going gets rough, turn to the books on style and grammar mentioned here, and to *Medical Style and Format.*[1]

The language of manuscripts can be copyedited either at the same time as their mechanical style and format (Chapters 5–7) or separately (Chapter 2). Language editing may also overlap with substantive editing (Chapter 3), but the items in Checklist 1 belong to copyediting proper and are the subject of this chapter.

Language editing has two main aspects: "editing for clarity and felicity of expression" and "matters of basic grammar, syntax, consistency and so on."[2] Aim to make or keep the author's language simple, clear, and unambiguous for the reader—transparent, in today's jargon. A functional rather than an elegant style is adequate in science, but if you meet a graceful style of writing, don't reduce it to the functional variety unless it is so precious that it distracts readers from the scientific message.

Whatever the style, don't make changes without a good reason, even if you think you are well qualified to do so: "A degree in English is not a license to violate the traditions, usages, nuances, vocabulary, and logic patterns of other fields," as Bernard Forscher[3] points out. Remember the Third Edict of Copyediting (Chapter 2), *Make essential changes only*, and its companions, *Find good authority for every change* and *Leave well enough alone*—but correct anything that is obviously wrong.

When you edit manuscripts by authors for whom English is a second language, aim to encourage rather than demoralize them. If you alter their English for apparently trivial reasons, or for no reason at all, they may lose confidence in their ability to write in English and the next manuscript

Checklist 1   Checklist for language

1. Spelling correct and consistent, according to journal preferences?
2. Punctuation correct?
   a. Quotation marks correctly used according to house style or national conventions?
   b. Parentheses correctly placed?
   c. Subjects and verbs separated by paired commas or none (never by one)?
   d. Commas in adjectival clauses correctly used?
   e. Hyphens correctly used? Included where they help the reader?
   f. Apostrophes kept to a minimum? Correctly used?
   g. Colons and semicolons correctly used?
   h. Exclamation marks and dashes kept to a minimum?
3. Grammar and syntax correct?
   a. Verbs agree with their subjects?
   b. Auxiliary verbs included where necessary?
   c. Participles, including infinitives, correctly attached to their subjects?
   d. Passive kept in its place (kept to a reasonable minimum)?
   e. Tenses correctly used?
   f. Abstract nouns used with restraint?
   g. Strings of three or more nouns kept to a minimum?
   h. Pronouns refer clearly to a preceding noun?
   i. Relative pronouns correctly used?
   j. Definite and indefinite articles and other particles included or omitted, as necessary?
   k. Prepositions correctly used?
   l. Comparatives complete?
   m. Words used in the right order?
4. Usage correct? Follows current recommendations?
   a. Words used with precision?
   b. The simplest words and the simplest ways of writing used?
   c. Sexist, racist, or dehumanizing terms transformed or removed?

they submit may be worse rather than better. Instead, keep alterations to a minimum and explain briefly why the less obvious changes are necessary. Be prepared to devote much more of your time to these authors than to native English-speakers, especially if their papers have been translated for them by a non-scientist.

You will soon learn which are the commonest mistakes made by the different language groups and be able to deal with them quickly. These mistakes include the use of definite or indefinite articles when they are not needed and their omission when they are needed, wrongly used prepositions, wrongly used tenses, wrongly used punctuation (commas in unconventional places), and wrongly used or redundant words (e.g., "respectively"). These and other faults commonly made in scientific writing by authors of all nationalities are described in this chapter.

(1) Spelling

When you first start work as a copyeditor, or when you go to work on a journal covering subjects new to you, look up the spelling and meaning of every technical word and every new word of any kind that you meet. This is time-consuming but rewarding, especially if you read the definitions as well as checking the spelling. After a while you will know not only how to spell the words but also a lot more about the subjects you are dealing with.

Some words have alternative spellings. Dictionaries usually indicate a preferred version, but sometimes two spellings are equally acceptable. Check how your journal's house style copes with words in this category. If house style is no help you'll have to decide between the dictionary's preference and the author's, if they are different. You may need to do a little research into why they are different. If the author is using the spelling preferred by others in the same discipline, don't change it unless the journal's house style or the editor is adamant on this.

Watch out for tricky plurals such as those listed below, and be sure that verbs agree with them:

Plural	*Singular*
phenomena	phenomenon
sera	serum
media	medium
criteria	criterion
symposia	symposium
genera	genus
analyses	analysis
matrixes (but matrices in mathematics)	matrix
indexes (but indices for measurable quantities)	index

Use English rather than Latin plurals for words that have become sufficiently anglicized (formulas rather than formulae), if house style allows this.

American and British spellings are another problem to be resolved. Some international journals allow authors to use either of these spellings, though not a mixture in the same paper. Other journals prefer the same brand of spelling to be used throughout, so you may need to change British spelling to American, or American to British, as appropriate. A dictionary that gives both American and British spellings will be useful (*The Random House College Dictionary*,[4] for example).

The main differences between American and British spellings (American given first) are in nouns ending in -er,-re; -se, -ce; -or, -our; -og, -ogue; -am, -amme (but "computer program" is correct in the United Kingdom). Then there are words with double or single consonants (labeled, labelled) and words formerly spelled with diphthongs, where American spelling uses "e" while British usually keeps "ae" or "oe" (but note that "fetus" is now widely accepted as correct in British medical journals, justification for this having been found in the *Shorter Oxford English Dictionary*).

Another difference between American and British spelling is in -ize and -ization. Many, but not all, British writers use -ise and -isation, in spite of the *Shorter Oxford English Dictionary's* pronouncement that

> The suffix . . . is in its origin the Greek *-izein*, L[atin] *-izare*; and, as the pronunciation is also with *z*, there is no reason why in English the special French spelling in *-iser* should ever be followed.

But, as the *Shorter Oxford* points out, -ise should be used in words not of Greek or Latin origin, such as advertise, advise, and some 40 others (see Ref. 5, p. 112). The *Shorter Oxford*, however, favors -yse endings (analyse, catalyse, dialyse, hydrolyse, paralyse) where the American preference is for -yze.

There are other differences in spelling too, far too many to list here (but see Ref. 5, p. 113–114). Again, look up every word of whose spelling you are not sure. And watch out for words that are often misspelled

> occured instead of occurred
> occurence or occurance instead of occurrence
> accomodation instead of accommodation

or that tend to be confused with words that look similar

> casual instead of causal

In quotations and in bibliographical references keep the same spelling or misspelling as in the original. Add "[*sic*]" if you want to tell readers that a word is being printed exactly as it appeared in the original.

(2) Punctuation

Make sure that manuscripts are correctly punctuated according to the general rules of punctuation (see Carey's *Mind the Stop,*[6] or a good book of English grammar, or Ref. 4, p. 106–111; Ref. 7, p. 66–91; Ref. 8, p. 125–138; Ref. 9, p. 132–155). Mistakes in punctuation are sometimes trivial but can sometimes change the whole meaning of a sentence. Whatever you do to the punctuation in the interests of readability, make sure you preserve the author's meaning. The commonest problems of punctuation in scientific writing are listed in (a) to (h) below.

(a) Incorrectly Placed Quotation Marks

American and British conventions for quotation marks (known as inverted commas in the United Kingdom) tend to differ. In the United States general usage requires double quotation marks for a direct quotation, with single quotation marks inside them for quotes within a quote. In the United Kingdom the more general usage is single quotation marks with double quotation marks inside them.

American style is to place closing quotation marks inside a semicolon or colon and outside a comma, a period, or any other punctuation marks that form part of a quotation; but quotation marks are put inside a question mark that is not part of the quotation.

British style is to use closing quotation marks outside an exclamation mark or a question mark, dash, or parenthesis belonging to the quotation. But if a complete sentence is quoted at the end of the main sentence, the closing quotation mark may go either inside or outside the period, depending on house style[5] (p. 196).

If more than one paragraph is being quoted (United States and United Kingdom), put opening quotation marks at the beginning of each paragraph but no closing marks except at the end of the last quoted paragraph, if the quotations are printed in the same way as the rest of the text. No quotation marks are needed for *displayed* material — material indented from the left margin or set in smaller type, or both. Quotations longer than eight or ten typed lines, or longer than one paragraph, are often displayed. Shorter quotations may also be displayed, depending on the emphasis called for by the text.

If words are left out of a quotation, use three spaced points (ellipsis points) to show where the words have been removed. If a period follows three ellipsis points circle it to show that it isn't just a surplus ellipsis point (but three ellipsis points on their own are also acceptable, not to mention more practical,[10] at the end of a sentence). Ellipsis points at the beginning and end of quotations are usually redundant; delete them unless the sense demands their presence[5] (p. 197).

On the whole, delete quotation marks (and italics) when these are used for emphasis or to alert readers to a novel or odd word or phrase. What the author thinks is unusual may in fact be familiar to readers, who are in any case presumably capable of noticing any oddities for themselves.

Use of quotation marks for emphasis or for oddities may occasionally be allowed at the first appearance of a word that doesn't have its usual meaning ("manuscript" where an electronic version is referred to, perhaps).

(b) Wrongly Placed Parentheses

If complete sentences are enclosed in parentheses, put the full point inside the closing paren. If words in the middle of a sentence are enclosed in parentheses, place any punctuation marks after the closing paren, not before the opening paren.

(c) Subjects Separated from Verbs by Single Commas

Never allow the subject of a sentence to be separated from its verb by a single comma. Instead, make sure there are either two commas or none:

WRONG   What Smith reported, was true. [Remove the single comma.]
RIGHT   Smith, as I said earlier, reported this result correctly.

(d) Wrongly Used Commas in Adjectival Phrases

The meaning of phrases describing the subject of a sentence can be changed by the presence or absence of commas. The message of the sentence

The fields, measuring about 400×600 meters, belong to X

is that all the fields under discussion belong to X. The words between the commas give the reader extra but not essential information about the fields. Without the commas the phrase is a restrictive appositive meaning that only the fields with those measurements belong to X, the implication being that there are other fields of a different size belonging to someone else.

(e) Wrongly Used or Missing Hyphens

Hyphens seem to be disappearing from the language, but they still perform a useful function. Do not abandon them totally unless your journal's house style insists on this. Omission in these examples was ill advised:

Random samples . . . were studied by means of a doctor administered questionnaire. [Hyphen needed between "doctor" and "administered."]

We believe that an activity mediated competition for a trophic factor then plays a significant role. [Hyphen needed between "activity" and "mediated."]

Hyphens are also useful for distinguishing between words such as "reform" (what some political systems need) and "re-form" (form again), "resign" (from your job) and "re-sign" (sign again), "unionized" and "un-ionized."

Depending on journal style or on what the approved dictionary in the subject recommends, you may need to insert or remove hyphens in words with prefixes ("semi-solid"), in compound terms ("copy-editor"), and between two vowels in words such as "re-election," "co-operation," and "microorganism" (see Ref. 9, p. 162–164, 176–181, for the principles of hyphenation).

Avoid floating hyphens: "the pre- and postoperative procedures were . . ." can easily be reworded.

Hyphens are rarely needed between adverbs and the words they qualify: "an effectively designed table" is correct, but "a little used car" is ambiguous.[11]

(f) Wrongly Used or Missing Apostrophes

The apostrophe is most often used to indicate possession ("nature's silyl group") or elision of a letter ("can't"). As both uses have an aura of slang, keep apostrophes to a minimum in scholarly writing. But leave them in when it would be pedantic to turn phrases round just to avoid an apostrophe ("the carboxylate ion is the silyl group of nature"), and put them in where the meaning is ambiguous without them (does "the authors manuscripts" refer to the manuscripts of several authors or to several manuscripts belonging to one author?).

The contraction "it's" ("it is") is often confused with the possessive "its" (no apostrophe). Sort out which is which and correct as necessary.

When words end in "s," add the apostrophe and a second "s" for the possessive ("James's") unless the word is a common noun followed by another noun beginning with "s" ("for peace' sake") (see Fowler[11] on "possessive puzzles" and "sake").

An apostrophe is sometimes used for the plurals of certain letters, figures, or words (P's and Q's, the '20's, why's and wherefore's) but is now more commonly left out (the 80s). Follow journal style in this.

(g) Confused Colons and Semicolons

Colons are used before lists, before clauses illustrating or providing a strong contrast or connection with the first part of the sentence, and before clauses introducing a climax or conclusion. They should not be used to introduce lists when the listed items themselves complete the sentence:

> For correcting manuscripts you need: a pencil, an eraser, and pens of different colors. [The colon should be removed.]

Semicolons are slightly weaker than colons. They are used to separate closely related clauses where the second explains the first or suggests a contrast with it, or describes a different aspect of the same topic; they are also used to separate parts of lists where the parts already include commas.

If an initial capital letter is used for the first word after a colon, consult your house style; this usage now seems to be common in the United States when an independent clause follows the colon[9] (p. 149-150) but it is not generally acceptable in the United Kingdom.

(h) Surplus Exclamation Marks and Dashes

Exclamation marks are rare in scientific writing, where their main use is as factorial symbols in mathematics or to indicate that botanical specimens have been examined by the author in person[8] (p. 165). Delete them elsewhere except when their presence in a particular sentence has some special justification.

Long dashes (em dashes) are used in almost the same way as parentheses, and sometimes as a third level of parenthetical matter, within square brackets that are within parentheses. Dashes emphasize the material they set off; parentheses indicate that the material is less important than the main topic of the sentence. If an author has scattered dashes around too freely, change them to parentheses, if appropriate, or reword the sentence.

The en dash, half the size of the em dash in the same typeface, is used to indicate distance or movement (London–Philadelphia), a combination (gas–liquid), or a range (50–60 ml). But use "from 18 to 21," not "from 18–21"; use "between 30 and 40," not "between 30–40"; and use "to" instead of an en dash when the dash could be confused with a minus sign. En dashes may also represent single bonds in chemistry.

(3) Common Faults in Grammar and Syntax

This is not the place to try to tell you all about basic grammar — see a good grammar book for that. Instead, as for punctuation, I'll outline the commonest faults in scientific writing, to alert you to the likely problems and to ways of dealing with them.

(a) Disagreement Between Verbs and Their Subjects

When verbs are a long way from their subjects or when there is a compound subject, the verbs sometimes end up disagreeing with their subjects in number. This mistake also happens when a plural noun does not end in -s (data) or when a singular noun does (kinetics). The verb in

> The proteolytic activity of extracts of X from these organs were expected to reach a high level

is wrong because the subject of the verb is "activity," not "extracts" or "organs." Check that plural subjects have plural verbs and that singular subjects have singular verbs.

If there are alternative subjects of unlike number, the verb agrees with the nearest element:

> Four rats or one dog was selected for this experiment.

But here it would be better to change the verb to "must be selected," or change the wording:

> Four rats were selected for this experiment, or one dog.

(b) Wrongly Used Auxiliary Verbs

Check that the various forms of the auxiliary verbs "to have" and "to be" have been included where necessary and omitted where they are not necessary in a series of passive verbs. In the sentence

> The sample has been weighed and several fractions taken for examination

the second verb needs a plural auxiliary instead of the implied singular.

> The sample has been weighed and several fractions were taken for examination.

Even when both subjects are plural, check that omission of the auxiliary does not make the second verb sound intransitive, as in

> The samples were weighed and four fractions examined

where it seems, to begin with, as if the fractions were doing the examining.

"There is," "There are," and so on are often overused or clumsily used. The sentence

> There were four experiments on X reported in the paper by Schmitt et al. (1980)

would be better as

> Schmitt et al. (1986) reported four experiments on X

— unless perhaps the writer wants to contrast the number four with some other number just mentioned or about to be mentioned.

(c) Detached Participles, Gerunds, and Infinitives

Participles are often left dangling (or hanging or unattached), or incorrectly attached, as in these two sentences:

> Before acquiring the grant the rats were housed in cages.
>
> Work on insect nervous systems designed to gain insights into Y.

Check that all participles, particularly those ending in -ing, are correctly attached to the nearest noun or pronoun. Put a fault right by substituting an active verb for the participle and naming the subject:

> Before we acquired the grant the rats were housed in cages.
>
> Our work on insect nervous systems was designed to gain insights into Y.

Similarly, gerunds (those verbal nouns ending in -ing) sometimes need a possessive noun or pronoun to make it clear who or what is the agent of the action:

> The inhabitants benefit by the architect's designing earthquake-proof buildings for the town.

Infinitives may become detached too:

> To do this work the insects were first weighed.

Check infinitives as well as past and present participles, and substitute the true subject where necessary:

> First we weighed the insects.

Avoidance of the first person pronoun is often to blame for participles being detached, and it contributes to the next problem too.

(d) Excessive Use of the Passive

The passive voice is correct and useful in scientific writing whenever the reader does not need to know who or what performed the action described by the verb ("The animals were fed three times a day"). The passive, however, tends to be overused by scientists, obscuring the agent of the action when the reader really needs to know who or what the agent is. The passive also produces a heavy plodding style ("It is thought that the excellent results obtained with this instrument were greatly facilitated by the care that was taken to calibrate it with model X7"). The commonest mistake is that authors, through mistaken modesty or careless writing, fail to differentiate between their own and other people's work in the discussion section of manuscripts. Where appropriate, change the passive to the active voice:

> We think that careful calibration of this instrument with model X7 was largely responsible for the excellent results we obtained.

Or, better still:

> The excellent results we obtained were largely due to careful calibration of this instrument with model X7.

But be discreet: do not change every single passive verb to an active one.

(e) Wrongly Used Tenses

The past tense is correctly used (1) for observations, (2) for completed actions, and (3) for specific conclusions:

(1) The insects weighed 0.5 g each.
(2) They reached the end of the road.
(3) The experiment was a success.

The present tense is used (1) for directions and (2) for generalizations or general statements:

(1) Put the camel through the needle's eye.
(2) No camel is small enough for that.
  This machine is the most successful of its kind.

Writers whose first language is not English sometimes confuse past and present tenses or the sequence of tenses, particularly in the various forms of the perfect ("I have lit the gas," "I had lit the gas," and so on) and the progressive or continuous tenses ("I am lighting the gas," "They have been gazing at the stars"). The sequence in this example is wrong:

> In transverse sections of these samples we often see some specific staining but it was a very small amount. [Make the tenses match by changing "was" to "is" or "see" to "saw."]

Correct the sequence of tenses when necessary, and try to make sure that the sequence agrees with what the author wants to say.

(f) Excessive Use of Abstract Nouns

As well as overuse of the passive, another sign of a plodding style is a surfeit of abstract nouns, particularly those ending in -tion, in company with "of," "the," and weak past participles (occurred, effected, brought about, achieved, produced) — as in this kind of clotted sentence:

> The formation of x without precipitation of y or z was effected by addition of a to the medium.

One sentence like that in a manuscript is more than enough. Instead, change the abstract nouns to active verbs and weed out the weak participles:

> When we added a to the medium, x formed without precipitating y or z.

(g) Overlong Strings of Nouns

Strings of nouns used as modifiers also help to knot up scientific prose. How long does it take you to decide what phrases like "adult rat muscle protein iron" mean? ("Protein iron in the muscle tissue of adult rats"?) You can save readers a lot of time if you insert hyphens or unravel phrases with three or more nouns in a row. But take care not to change the meaning when you disentangle the nouns. And don't apply this rule of three overenthusiastically. Two nouns, or sometimes three or four ("Norwalk serum antibody"; "reciprocal geometric mean titer"), often constitute a single name in science; count these as one compound term when you count the number of words in a string with a view to unraveling it. Then there will be occasions when discretion is the better part of valor and you might leave the author's version undisturbed, provided the readers are likely to understand it; this is another of those moments for applying common sense.

(h) Unattached Pronouns

Pronouns must refer unambiguously to preceding nouns, but "it," "this," "that," and "which" often lose their antecedents. Check that all pronouns, particularly "it" and "this" at the beginning of sentences, relate clearly to earlier nouns, phrases, or complete sentences.

> Current efforts are aimed at exploring the possibility that some of these defects may account for the hallmarks of malignancy. This is hampered by a scarcity of information on the natural history of human cancer.

If "This" in the second sentence of that example refers to the exploration, it would be better to include the word "exploration." The next example is more difficult to deal with:

In older embryos, however, coupling is more restricted and dye injected into ICM cells spreads throughout the ICM but not into the surrounding trophoblast. This is not a boundary, however, between two coupled compartments.

Where the antecedent is not clear, ask the author to clarify the sentence ("This = what?"), or substitute the noun or phrase you think the pronoun refers to and ask the author to confirm or correct your suggestion.

(i) Imprecise Use of Relative Pronouns

"That" and "which," strictly speaking, have distinct meanings when they introduce defining or non-defining (or restrictive/non-restrictive) adjectival clauses. For defining clauses, such as

The powder which resulted from this treatment was too fine

the recommended usage is "that" in place of "which":

The powder that resulted . . .

When "which" is used with a pair of commas (see section 2d above), the sentence becomes a non-defining one giving extra but not essential information:

The powder, which resulted from this treatment, was too fine.

However, "which" is very commonly used instead of "that" in sentences like the first example here. It would be pedantic to go on a witch hunt for "whiches" unless your journal is adamant about making the distinction. Instead, change "which" to "that" (or vice versa) only when the distinction seems to be useful, or when the use of "which" avoids too many "thats" in a sentence.

(j) Incorrect Use of Articles and Particles

The definite article "the" and the indefinite articles "a" and "an" are sometimes missing where they are needed, or included where they are not needed, especially in papers by authors for whom English is not the first language. Insert articles before common nouns in the singular ("He lit *a* cigarette") except when the common nouns are titles preceding proper nouns ("Mary spoke to *Professor* Smith"). "Mary spoke to *a* Professor Smith" would be correct only if the author wanted to imply that Professor Smith was someone the readers were unlikely to have heard of before.

Use the definite article when a common noun is particularized ("He lit *the* fire for her") and an indefinite article when the noun is generalized ("*A* cigarette is made of tobacco rolled in a tube of paper"). If the common noun is in the plural, remove the article ("Cigarettes can harm your health") except when the noun is particularized ("*The* cigarettes in that packet taste awful"). Also remove articles before proper, material, or abstract nouns ("London," "iron," "movement") except when these are being used as common nouns (as they are in "The London she knew 50 years ago was a different city," "The iron in those rails," and "The movement was a sudden one"). But insert or keep articles where proper nouns are the names of rivers, seas, or oceans ("the Delaware," "the Pacific"), or of groups of islands or ranges of mountains ("the Aleutians," "the Himalayas"), or where the noun is a book such as "the Bible."

Base your choice of "a" or "an" on the accepted pronunciation of the first syllable of a word or abbreviation: "an RNA"; "an herb" (in the United States, where the "h" is not sounded) but "a herb" (United Kingdom) and "a hotel" (because the "h" is sounded). Minor dilemmas will appear with abbreviations, though: some people think of mRNA as "messenger RNA" and of NaCl as "sodium chloride," while others pronounce them mentally as "em RNA" and "en a see el" — I think the latter is to be preferred, with "an."

A surfeit of negatives, whether these are particles such as "not" or "nor" or other kinds of negative words, can make sentences either wrong or difficult to understand:

> Only in this laboratory was there insufficient equipment to prevent the work from being completed.

Two negatives usually add up to a positive, in grammar as in arithmetic, so reword sentences with negative constructions whenever they could confuse readers.

> Even then, however, there will still be many arguments in favor of not storing documentary or other high-resolution images in digital form but in direct optical form

is easier to understand at first reading if it becomes

> Even then, however, there will still be many arguments in favor of storing documentary or other high-resolution images in direct optical form rather than in digital form.

(k) Incorrect Use of Prepositions

Prepositions in the English language can cause a lot of minor and a few major problems for native English speakers as well as for authors whose first language is not English. Check in a good dictionary or grammar book that words are followed by their usual prepositions ("I am tired of," not "I am tired with"), that prepositional phrases are correctly used, and that prepositions are included where they are needed and omitted where they are not needed.

A common fault is using "different than" instead of "different from" or "different to." Allow "different than" only in sentences such as "The laboratory has quite a different appearance now than it had last time I saw it," where it does away with the pedantic "different . . . from that which" or with the ugly and incorrect "different . . . to [or from] what it had."

Strict application of the old rule that prepositions should not appear at the end of sentences is pedantic. Unwind sentences that authors twisted in their efforts to avoid this supposed fault, as in

> The thickness of the slices depends on the type of knife with which we cut them.

This sentence would sound more natural as

> The thickness of the slices depends on the type of knife we cut them with.

(l) Incomplete Comparatives

Watch out for incomplete comparatives. If an author writes "This method was more difficult," or "simpler," and stops there, is it clear what the method was more difficult or simpler than? Write the missing term in, if you can, or ask the author to supply it.

(m) Incorrect Order of Words

Incorrectly ordered words lead to ambiguity, as in

> John Smith, pictured (right) with his son, aged 5, who received the Medal for Gallantry yesterday

where the final clause should have followed the name if John Smith, rather than his young son, received the medal.

Check that words are placed in their natural order, which in English sentences is most commonly (but not invariably) subject, verb, object:

> The participants presented their lectures in the mornings

not

> Their lectures the participants presented in the mornings.

Watch in particular for unnecessarily split infinitives:

> These unedited papers bypass technical editing in order to more quickly and economically disseminate information.

Occasionally, however, it is better to split an adverb from a closely linked infinitive than to allow ambiguity or artificiality:

> They planned to quickly reassemble essential parts taken from the original machine

is preferable to the alternatives. And remember that an auxiliary verb can quite properly be separated from the main verb (participle or infinitive) in a sentence such as "They wanted the parts to be carefully matched."

Adverbs or adverbial phrases often go immediately *before* the words they qualify, but they go *after* intransitive verbs ("The temperature fell quickly") and they must not separate transitive verbs from their objects (as "within 30 seconds" does in "The sample was replaced within 30 seconds in the analyzer").

A word that often ends up in the wrong place is "only," which can be used as either an adjective ("Only she knew where the key was kept") or an adverb ("She knew only where the key was kept"). Make sure that the position of "only" gives the intended meaning, as far as you can determine it, but do not change it if the meaning is clear even though the position is technically wrong: "He only died a week ago," as Fowler points out,[11] is just as clear as "He died only a week ago," which has "only" in the orthodox position.

(4) Usage

Make sure that words are used correctly and to their best effect, and that current recommendations on usage are followed. (See Ref. 8, p. 269–278, for a list of confusing pairs and a list of words that are often misused or misspelled.)

(a) Precision

Has the author chosen something that resembles the intended word but has a different meaning? In

> The obvious step was to *disperse* with the carrier

and

This should not *detract* from the task of characterizing Y

"dispense" and "distract us" should replace the italicized words. Correct obvious errors of this kind and query any words or phrases you are not sure about. Here again, as with spelling, you should look up unfamiliar words if you suspect they have not been properly used.

Mixed or wrongly used metaphors or similar phrases are other items to look out for:

The patient had large hands to boot.

When possible, substitute more precise and concrete wording, appropriate to the context, for vague words such as "area," "character," "conditions," "field," "level," "nature," "problem," "process," "structure," and "system." "In the well-irrigated situation" is more precise and informative if it becomes "In well-irrigated fields" (where "fields" has its literal sense).

Delete any unnecessary adjectives or adverbs, especially vague qualifiers such as "very," "quite," "rather," "fairly," "relatively," "comparatively," "several," and "much"—or ask the author to be more specific, where necessary.

Convert euphemisms such as "sacrifice" or "passed on" into the words they hide ("kill," "die") where journal practice allows this.

The mice were sacrificed on day 10 of the experiment

would be better as

The mice were humanely killed on day 10 of the experiment.

Change "in the case of" to "in," "for," or another appropriate preposition, depending on the context. Change "following" ("Following affinity chromatography, the protein was sequenced") to "after" and "prior to" to "before," except when "following" and "prior" are used as adjectives ("the following day," "a prior engagement").

Banish "recent" and "recently" ("Recent work by Schmitt [1980] shows") when they refer to work published more than a year before the publication date of the manuscript you are copyediting.

Remove any slang, jargon, or phraseology more suited to conversation between scientists in the same discipline than to the printed page. Write "laboratory" in full when "lab" appears, and change terms such as "euthanized," "melanized," and suchlike to words that appear in dictionaries. In biomedical writing "pathology," "morphology," and "etiology" are constantly used to mean "abnormality," "structure," and "cause," although their real meaning is the study or philosophy of disease, form, and causation.

Change these words and their siblings to more precise terms when necessary. (Recognizing slang or unnecessary jargon may not be easy at the start of your copyediting career, but you could perhaps consult your editor or a more experienced colleague.)

The word "data" is often abused by being used to refer to "results" (what was found) when its real meaning is "facts or information, esp. as basis for inference" (*Concise Oxford English Dictionary*). That is, "data" refers to the starting-point for a hypothesis, for example, not to the end-point of the experiments that aim to test the truth of that hypothesis. Is it too late to insist on the correct use of "data"? Consult your journal's house style on this and on whether to treat the word as a singular or — more correctly — as a plural noun.

Check that sentences containing "so that" are unambiguous.

> The females of this species are poor providers so that many of their offspring die young

implies that the mothers intend to kill off their progeny. Turn the sentence around:

> Many of the offspring of this species die young because the females are poor providers.

"Because" becomes ambiguous when it follows a negative clause:

> They did not report the experiment because they hoped to win a Nobel Prize

could mean either that they reported the experiment, though not because they hoped to win the prize, or that they suppressed the report because it would ruin their chances of the prize.

Correct sentences in which the subject has been switched half-way through, leading to confusion, as in

> Bioactive silicates with less than critical concentrations of such multivalent cations show bonding with osteoblasts, produce matrix vesicles, and normal mineralization proceeds at the interface.

(b) Simplicity and Comprehensibility

The general principle in scientific copyediting, as already mentioned, is that authors' prose should be changed only when the grammar is incorrect or when the style makes it difficult for readers to understand the message. For example, use your judgment about changing elaborate words for which simple alternatives exist. If the meaning is already clear, it may be better to preserve the author's good will until you need approval for a more important change. But substitute simpler Anglo-Saxon words when authors persistently use polysyllabic or archaic words (change "terminate" to "end" and "pertaining to" to "on," for example) and simplify or remove pompous phrases: "equivalent as far as acceptability is concerned" can become "equally acceptable," "audible distress vocalization" means "crying," and "Restrictions on ambulation are often favorable to inhibition of edema" tells us that "Rest often inhibits edema."[12]

Remove phrases and sentences that obstruct comprehension because they say nothing or almost nothing:

> At the end of the day we shall hopefully have achieved significant insights into the ongoing condition in which these symptoms have manifested themselves.

Style manuals and other books on writing (see Ref. 8, p. 36–37; Ref. 13, p. 93–98) have useful lists of commonly used phrases that can be simplified or omitted. When you remove redundant words or phrases ("the nervous systems of insects *share* properties *in common* with other animals"; "It is worthy of note that"), you save money for the publisher and time for the reader, and you help to make the author's message easier for the reader to grasp.

Legends or captions, often written as if illustrations were invisible, are a favorite home for redundant phrases. Remove "Diagram of," "Photograph of," "Map of," and so on.

Hedging produces more unnecessary verbiage. Watch out for words such as "suppose," "indicate," "appear," "seem," "may," "feel," and so on.

> It would seem reasonable to suggest that the X effect may possibly be due to Y

means no more than

> The X effect may be due to Y.

One degree of hedging ("may" in the corrected example) is usually enough, although two may sometimes be allowed for emphasis. Correct or query a retreat into hedging after a positive phrase:

> We firmly believe that this substance may play a part in the X effect.

(c) Sexism, Racism, and Dehumanizing Terms

Transform sexist terms to non-sexist versions whenever you find the former. Avoid overuse of "he or she"; instead, change sentences to the plural. "The scientist is not always a good writer; he needs to attend courses in writing" can become "Scientists are not always good writers; they need to attend courses in writing." Substitute neutral words for words that refer only to the male sex (except of course where the person referred to is indeed a man). This kind of substitution is an aid to precision, particularly in medical writing, where phrases such as "Breastfeeding in man is . . ." are not uncommon. (See the American Psychological Association's Publication Manual[14] or Miller and Swift[15] for more advice on changing sexist language.)

Racist or chauvinist expressions are rare in scientific writing. If you come across any, remove them. In medical papers remove "Caucasian" or "black" and so on if race has no clinical implications. Where the name of a race is used, make sure that it is both acceptable and correct. "Native American" now seems to be preferred to "American Indian"; "Oriental" should not be used to include Asian Indians.

Dehumanizing terms are common in medicine, where people are often referred to as "subjects" or "cases" or, for example, as "amnesiacs," "geriatrics," or even "pediatrics." Change these terms to "volunteers," "patients," "patients with amnesia," "patients in geriatric wards," "sick children," and so on, as appropriate.

Finally, another reminder — make all the changes in grammar and style that are necessary, but none that are not: *Leave well enough alone.*

References

1. Huth EJ.   1986   Medical style and format: an international manual for authors, editors, and publishers. ISI Press, Philadelphia, in press.
2. Bell JG.   1983   On being an uncompromising editor. Scholarly Publishing 1983; 14:155–161.
3. Forscher BK.   1985   Preferred background for manuscript editors: English or science? CBE Views 1985; 8(3):5–7.
4. Random House college dictionary, revised ed.   1982   Random House, New York, 1982.
5. Butcher J.   1981   Copy-editing: the Cambridge handbook, 2nd ed. Cambridge University Press, Cambridge, 1981.
6. Carey GV.   1971   Mind the stop: brief guide to punctuation. Penguin, Harmondsworth, UK, 1971.
7. Judd K.   1982   Copyediting: a practical guide. Kaufmann, Los Altos, CA, 1982.
8. CBE Style Manual Committee.   1983   CBE style manual: a guide for authors, editors, and publishers in the biological sciences, 5th ed. Council of Biology Editors, Bethesda, MD, 1983.
9. University of Chicago Press.   1982   The Chicago manual of style, 13th ed. University of Chicago Press, Chicago, 1982.
10. Harman E.   1976   A reconsideration of manuscript editing. Scholarly Publishing 1976; 7:146–156.

11. Fowler HW.   1965   A dictionary of modern English usage, 2nd ed. revised by Sir Ernest Gowers. Oxford University Press, Oxford, 1965.
12. Anon.   1976   Lancet 1976; 2:1120.
13. O'Connor M, Woodford FP.   1975   Writing scientific papers in English: an ELSE–Ciba Foundation guide for authors. Excerpta Medica, Amsterdam, 1975/Pitman, London, 1978 [2nd edition, Wiley, Chichester and New York, 1987].
14. American Psychological Association.   1983   Publication manual of the American Psychological Association, 3rd ed. American Psychological Association, Washington, DC, 1983.
15. Miller C, Swift K.   1980   The handbook of non-sexist writing. Lippincott & Crowell, New York, 1980/Women's Press, London, 1981.

Chapter 5

Technical Editing (Mechanical Style) — Text

Mechanical style and format, which are part of technical editing (Fig. 4, p. 4), can be dealt with at the same time as language or during a separate reading of the manuscript. Mechanical style and format can also be dealt with separately, perhaps by different people (see Van Buren and Buehler[1]), although many copyeditors work on them simultaneously. Some aspects of format are discussed with mechanical style in this chapter and in Chapter 6. Mark-up, the main part of formatting, is covered in Chapter 7.

Whichever method of working you choose, take care of all the items listed in Checklist 2. Don't be put off by the length of this list; it should take less time to do most of these things than to read about them. (See Chapter 2 for advice or reminders on how to mark manuscripts.)

(1) Completeness and Length of Manuscript

Check first that the manuscript you receive (or that is about to be submitted, if you are an author's editor) still has all its vital parts — title page, abstract, text pages, and so on. Check that at the logging-in stage the number of tables and figures was noted on the title page (circle the information to show that it is not to be typeset) and that these details match what is recorded on the journal's chart or computer file.

You may be asked to record the number of characters or words as well as the number of tables and figures. If so, count the number of characters, including spaces, in an average line. Divide the total by 6 to get the average number of words per line (in English, the average word — admittedly including the non-scientific variety — consists of five letters and one space).

Checklist 2 Checklist for copyediting mechanical style and format (text)

1. Manuscript complete? Text pages consecutively numbered?
2. Reference number assigned to the manuscript?
3. Title page complete and properly marked?
4. Abstract included (if the journal requires one)?
5. Key words and bibliographical information included (if required)?
6. Text legible? Spelling and punctuation consistent?
7. Headings and subheadings in the text clearly ranked?
8. Indention or its absence indicated?
9. Capital letters, small capitals, italics, and bold type clearly marked and used consistently?
10. Abbreviations and symbols used correctly, and abbreviations explained at first mention?
11. Greek or Cyrillic letters and special scripts and accents clearly marked and consistently used?
12. Numerals correctly written and marked? Mathematical and chemical terms and equations correctly written?
13. Parentheses, brackets, and mathematical braces properly paired off?
14. Spacing between mathematical operators, numerals, and so on marked appropriately?
15. Acknowledgments properly positioned and marked?
16. Appendix included? Properly positioned and marked?
17. Footnotes and endnotes (if any) clearly linked to their intended positions?

Multiply the number of characters or words by the number of lines per page and the number of text pages. Make allowances for pages shorter or longer than the average, or total each page separately for greater accuracy. Make a separate list of character or word counts for references, figure legends, or any other sections likely to be set in a size of type different from the text type, or set separately from the text and from each other.

Make sure that manuscript pages (also known as folios) have been numbered consecutively, preferably in the top right-hand corner. If any pages have been inserted after page 8, for example, renumber page 8 as 8A and number the inserted pages 8B, 8C, and so on (or use the sequence 8, 8A, 8B, etc., if this is house style). For total assurance that no pages have been left out you might also write "p. 9 follows" at the bottom of the last inserted page and "p. 8A and 8B precede" at the top of page 9. If a page has been removed or the numbering has jumped a page, renumber the page that comes before or after the gap: if there is no page 5, for example, change page 4 to "4 and 5," or page 6 to "5 and 6."

(2) Manuscript Reference Number

Check that a reference number was indeed provided when the manuscript first arrived in the editorial office. If necessary, assign a number now. Write the number, or make sure it has already been written, on the title page and on any parts of the manuscript — such as the reference list, tables, figures, legends, and footnotes — that may be handled separately from the rest during the copyediting or production processes.

(3) Title Page

The title page of a manuscript includes the title of the article and, usually, the byline, i.e., the names and addresses of the authors (Fig. 8). If other information is included that is not to be printed at the head of the article — such as a short title for page headlines ("running heads") or footlines, the address for correspondence, or the address to which reprints should go — circle this information. The instructions to authors and the example of a recent issue of the journal should cause the title-page items to be provided in the most convenient way for typesetting. If they are not in the right order, show the typesetter how they should be rearranged to match journal style (Fig. 8). If the amount of redundant material on the title page or on any other page is likely to distract the typesetter, blank out the unwanted parts completely, using correcting fluid or strips of adhesive paper — but keep a photocopy of the original in case a query arises later.

If you are an authors' editor, prepare the title page in the required form before the manuscript is submitted to the target journal. Show whether the title and the byline are to be centered, or start at the left margin (ranged left), or end at the right margin (ranged right). See the list of manuscript marks in Table 1 (p. 25) for the appropriate instructions.

When necessary, show which author works where by using superscript symbols or letters or numbers to link names to addresses (A. Lumpkin[a], B. Learner[b] . . . [a]Dept of Chemistry and [b]Dept of Biochemistry, University of North Yorkshire). If the addresses or other details are to appear as a footnote on the first page of the article (perhaps also linked to the names by a superscript symbol or letter), write "footnote" (circled) beside the line or lines containing these details.

Also mark letters or words that are to be printed in capitals, small capitals, italics (sloping), or bold (heavy) lettering. Is the whole title to be in capital letters (three underscores), or are all main words to have an initial cap? Or is an initial cap to be used only for the first word and for others normally written with an initial cap? How is the byline to be printed? Small caps (two underscores) are sometimes used for authors' names, for example. Italics (one underscore) may be used for items such as authors' addresses (see also section 9 below.)

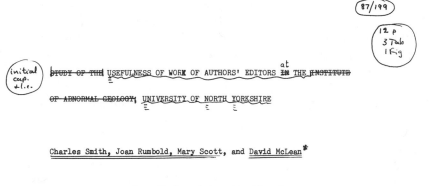

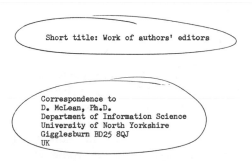

Short title: Work of authors' editors

Correspondence to
D. McLean, Ph.D.
Department of Information Science
University of North Yorkshire
Gigglesburn BD25 8QJ
UK

Figure 8   Sample title page, edited and marked up for typesetting.

If the typesetter is following a standard set of specifications for the journal, you may simply need to key the various parts of the title page to match the specifications (see Chapter 7).

If it is your responsibility to see that the Copyright Clearance Center's code or other codes are provided for the typesetter, write them on the title page if they are needed at the beginning of the article, and show where they are to be printed (see Chapter 8).

(4) Abstract

An abstract or summary at the beginning of an article is often set in a type face or type size different from the rest of the article, and it may be indented on the left, or set across two or more columns, or vary in some other way from the main text. Mark this section for the typesetter by giving the column width in picas (see Chapter 7) and stating the type face, size, and leading (e.g., Times Roman 8/9). Add the word "Abstract" at the beginning if that is the publication's practice, and mark its position and typographical style (italic or bold?) as appropriate.

(5) Key Words and Bibliographical Information ("Biblid")

If key words are added after the abstract, show whether they are to start on the last line of the abstract or on a new line immediately below the abstract, or go anywhere else. Underscore them if they are to be printed in italics.

If your journal prints a line or two of bibliographical information (biblid: see ISO/DIS 9115[2] and Chapter 8) before or after the abstract, or as a footline or elsewhere on the title page, mark it for the typesetter according to the journal's usual typographical style: roman type or italics, with or without a line space between it and the abstract.

(6) Legibility and Consistency

As you work through the manuscript, make any illegible words, symbols, or numerals legible, mark hyphens typed at line-ends to show whether they are to be kept in or deleted, and remove any unwanted underlining or any other now-redundant marks and messages from the author that could baffle or delay the typesetter or typist. Mark superscript and subscript letters or numerals. Where any doubt could arise, show the typesetter when "l" is a letter ("el") and when it is the numeral 1 (one), when "0" is a capital letter and when it is a zero, and when "x" is a multiplication sign rather than a letter of the alphabet. Mark all en dashes, em dashes, and minus signs for the typesetter; if you fail to do this, all dashes may be printed as hyphens.

Check whether spelling and punctuation are consistent (see also Chapter 4).

(7) Headings in Text

The main text of manuscripts often starts with the heading "Introduction." Is it journal practice to keep this in? If so, should it be centered or flush left, set in boldface or italic, and in capital letters throughout, or with an initial capital, with the rest in lowercase ("initial cap and l.c.")? Should other headings have an initial cap only or should they have initial caps for most words ("u. & l.c.")? A typical pattern of headings where four levels are needed might be:

FIRST-ORDER HEADING IN BOLD CAPITALS

Second-order Subheading in Bold u. & l.c.

Third-order subheading in italic initial cap & l.c.

Fourth-order subheading, a paragraph lead-in, italic initial cap & l.c. The text continues on the same line as this fourth-order subheading.

If there are more than four orders of headings and subheadings, one or more of them may be centered. Usually two or three orders are enough.

Key the headings for the typesetter by writing A, B, C, or D (or 1, 2, 3, 4) in the margin according to an agreed system and circle these letters or numbers, as for other instructions. Or mark up the headings fully, including the type size, as necessary.

If headings are numbered or lettered, what kind of numbers or letters are used in the journal? And are the numbers or letters placed between parentheses, or followed by a paren or a period?

(8) Indention

The first line of the first paragraph of an article and the first line of the first paragraph after any flush-left heading or subheading (starting at the left margin) may be printed either flush left or indented by whatever amount is standard for other paragraphs (usually one en or one em). If paragraphs in the manuscript have been indented when they ought to be flush left, mark them for the typesetter. Alternatively, if paragraphs or other parts that have been typed flush left are to be indented, mark the indent and its size.

Some paragraphs may be centered or displayed (indented from the right margin as well as the left), perhaps to set off a list or a quotation from the rest of the text. Mark these on both sides and confirm the instruction by writing "center" or "display" in the margin. Paragraphs of this kind take up more space than unindented ("full-out") text, so don't use them if they aren't really necessary. It is often enough to indent only the first line of each of a numbered series of points or paragraphs, or to indent on the left only.

Numbered paragraphs and entries in reference lists may be set "flush and hanging," that is with the first line flush left and the rest indented.

(9) Capitals, Small Capitals, Italics, and Boldface

Initial capitals (caps) are used for the first words of sentences and for proper names, titles, and trade names. Caps or initial caps are also used for chemical and other symbols, for the scientific names of phyla, orders, classes, families, or genera, and for many other kinds of scientific terms. If you are in doubt about scientific usage, consult the journal's house style, the *CBE Style Manual*[3] (p. 95–99), or a manual in the appropriate discipline.

If caps have been typed in the places where they are needed, don't underline them unless any doubt might arise — as it might, for example, when a proper name is included in a title or heading that is typed in caps but is keyed for setting in an initial cap and lowercase letters. In this case, underline (three times) any initial letter that should remain a cap. And if lowercase letters have been typed instead of caps, underline those letters three times.

Small caps are used occasionally for design purposes, and they are used for a few abbreviations or symbols (see Ref. 3, p. 96, p. 207–210; Ref. 4, p. 32). Double-underline any words or letters that are to be printed as small caps.

Italics are used for many different symbols, constants, letters, numbers, and names in science. Anything underlined once in the manuscript will be printed in italics, so add or delete underlining as necessary. Consult the *CBE Style Manual*[3] (p. 95–96) for scientific usage and the *Chicago Manual of Style*[5] or Hart[6] for other uses.

Remove underscoring of words when authors have used this for emphasis more than once every five typed pages. The italicization that will appear in the printed article loses its point if there is too much of it.

Bold type is used for certain names, vectors, and symbols (see Ref. 3, p. 99). Add or remove the wavy underlining that indicates bold type, as necessary.

(10) Abbreviations and Symbols

Abbreviations, including acronyms formed from the initial letters of compound terms, are short versions of words or phrases. Symbols are characters or marks that are accepted as the conventional signs representing objects, ideas, or processes.

Système International (SI) units for length, mass, time (m for meter, g for gram, s for second), and so on, and other abbreviations accepted in the discipline or more widely, need not be written in full when used with numbers. Most other abbreviations need explanations. Write every abbreviated term in full at first mention and give the abbreviation after it in parentheses. Ask the author to write terms in full if you can't discover what abbreviations stand for. Consult your journal's house style, the *CBE Style Manual*[3] (p. 244–255), or a manual for the relevant discipline to see when abbreviations should be used and which of them can stand on their own without being explained.

If the journal explains all abbreviations in a footnote or box near the beginning of each article, an alphabetical list of abbreviations, with their explanations, should form part of the manuscript. Check that the list is complete and matches the usage in the text. Compile the list yourself if the author has not supplied it.

Once an abbreviation has been explained, make sure it is used every time it is needed: if the author switches back to the full term, replace it with the abbreviation. On the other hand, if there are too many abbreviations, making sentences almost unreadable, write some of the terms in full every time they appear. Authors who use abbreviations extravagantly need to be restrained. For the sake of readers, decode sentences such as

MPTP is converted by MAO-B to MPP$^+$, which reaches SNpc nerve cells via DA uptake systems.

Reduce the strain on readers' memories by reducing the number of abbreviations in each manuscript to four or five, at most, in addition to the agreed SI abbreviations for units of measure. If an abbreviation is used only two or three times in a manuscript, write out the full term each time. Keep abbreviations to a minimum by substituting appropriate pronouns or nouns:

RNase was . . .

might be changed to

The enzyme was . . .

if RNase has already been mentioned, but be careful that this kind of change does not produce ambiguity.

Note that the international rules of bacteriological nomenclature allow the name of a genus to be abbreviated after the full name has been given at the first mention (*Salmonella typhimurium* becomes *S. typhimurium*, *Drosophila melanogaster* becomes *D. melanogaster*). If another species with the same initial letter for the genus is discussed in the same article, use a different abbreviation for the second species (*Staph.* for *Staphylococcus* if *S.* is used for *Salmonella*), or write the names in full every time if the journal disapproves of abbreviations of this kind.

Some abbreviations, especially acronyms, are always printed in capital letters, others (fewer) are always printed in lowercase letters, and yet others may be in either caps or lowercase, depending on disciplinary or national preferences. Abbreviations printed in caps rarely have punctuation or spaces between the letters (EMG). Lowercase abbreviations, however, often have full points after the letters (e.m.g.).

Units of measure are usually abbreviated only when they are used with numbers:

The sample weighed 5 g.

How many grams does it weigh?

If abbreviations for such units have periods after them in a manuscript, delete these, and delete -s if it has been added to show the plural (mgs. becomes mg). Depending on house style, you may also need to remove periods if these have been used with contractions — abbreviations that end in the same letter as the full term (Dr, not Dr.).

Check that symbols are the standard symbols accepted in science as a whole or recognized in a discipline. If they are unique to a particular manuscript, ask the author to explain them.

(11) Greek or Cyrillic Characters, Other Special Scripts, and Accents

Greek letters are used in the nomenclature of many branches of science (see *CBE Style Manual*[3]). Identify these in the margin at their first appearance ("l.c. Greek alpha," circled) and make sure that handwritten characters are clear to the typesetter. If Greek letters have been spelled out ("alpha") where normal practice in the discipline is to use the character itself ("α"), substitute the character. Do the same for Cyrillic characters and any other special scripts in the manuscript.

Call the typesetter's attention to unusual accents (diacritical marks) — for example the stod, the inverted cedilla, and the stroke (see Ref. 3, p. 138, 139) — and to unusual letters such as the undotted i. Write these accents or letters, or their names or descriptions, in the margins and circle them. Make a list of them for the typesetter if this is journal practice.

(12) Numerals and Equations

Use numerals for the day and year (1 May 1986) or for the year, month, and day (1986-05-01) when dates are cited. Do not use 5.1.86, which means 5 January in the United Kingdom or May 1, 1986, in the United States — unless house style stipulates this or a similar form.

Use numerals for times (8:15 p.m., 20.15, or 2015, depending on house style), and for page numbers and decimal quantities. Also use numerals for percentages (5 percent, 5 per cent, or 5%, according to house style), for quantities attached to standard units of measure (10 kg), and in other arithmetical or numerical contexts ("multiplied by 9"). Elsewhere follow house style — which may require words to be used for numbers one to nine (or ten, or twelve, or ninety-nine), and at the beginning of sentences. Where numerals are used for 10 and above (for example), use numerals for all numbers referring to items in the same category:

> He ordered 9 red, 18 yellow, and 25 green tiles.
>
> He needed 6–12 bricks.

(See Ref. 3, p. 146–147; Ref. 4, p. 141–149; Ref. 7, p. 99–102.)

In numbered lists or headings (see section 7 above) parentheses or periods are usually used to set the numbers off from the wording that follows. Mark these according to journal style. If journal style is flexible, use the author's style if it is consistent; otherwise impose a consistent style of your own.

In mathematical equations underline (once) the letters standing for variables ($x + y = z$) and for physical constants that are usually printed in italics. Identify for the typesetter any zeros, ones, and multiplication signs that could be mistaken for the letters O, l, or x (oh, el, ex) and point out prime signs ($'$ or $''$) and other symbols, and Greek and other such letters (write the name of the symbol or letter, circled, in the margin). If displayed (centered) equations are likely to overflow the printed line, show where they may be legitimately separated (preferably before a sign of operation [$+$, $-$, $\times$, $\div$] or after a logical grouping). Number a sequence of equations at the right-hand margin (usually), between parentheses. For equations written in the text line, many journals prefer to transform expressions to allow typesetting on a single line (cheaper and more aesthetic than having oddly spaced lines in the text), or on as few lines as possible. When house style favors transformation, change $\frac{a + b}{x + y}$ to (a + b)/(x + y) or use "exp" to change $Ae^{\frac{1}{2}(Vo/D)}$ to A exp 0.5(Vo/D) (example borrowed from Ref. 3, p. 29; see Ref. 7, p. 224–232).

When superior letters or numbers appear with inferiors, does the journal prefer $A_b{}^2$, A_b^2, or $A^2{}_b$? The second or "over and under" style is neater than the first, as Butcher points out[7] (p. 227), while the third style is better used with parentheses: $(A^2)_b$.

Use one solidus only; two produce ambiguity. Change "10 J/g/s" to "10 J/(g s)" or "10 J g^{-1} s^{-1}."[8] If the space between units in these examples is unacceptable, use a raised or centered period ("medial multiplication point") to indicate multiplication[7] (p. 230).

Decimal points should be on the line, not raised. Authors from some countries tend to use decimal commas: change these to full points. The point should always have a number in front of it (0.1, not .1), to reduce the chance that it will be overlooked by the typesetter or the reader.

In chemistry, as in mathematics, identify ambiguous numbers, name Greek and other letters, and name unusual symbols at their first appearance. If a chemical symbol such as He or I (for helium or iodine) could be mistaken for a word at the beginning of a sentence or elsewhere, avoid using the symbol (see Ref. 3, p. 216). In chemical formulas, omit bonds if it is not essential to show linkages (as in H_2O). Mark any bonds used in the text by writing "bond" against them if the typesetter has this special character (or "sort") available. Mark a medial or superior dot for free radicals (H· or H˙) and a double dot for lone pairs in an appropriate place above, below, or beside the symbol they belong to (N:, N̈:).

For structural formulas, authors should supply professionally prepared drawings. If the versions you receive have to be redrawn, make any necessary editorial changes on the originals or on photocopies before the originals go to the artist. Send photocopies of the redrawn structures, not the new drawings themselves, to the author for checking.

For other advice on copyediting in chemistry/biochemistry see the *CBE Style Manual*[3] (p. 214–226), Butcher[7] (p. 234–237), and *The ACS Style Guide.*[9] See the most recent editions of the *Merck Index*[10] or Marler[11] or the *Handbook of Chemistry and Physics*[12] for correct chemical names and formulas.

(13) Parentheses and Brackets

Check that parentheses, ( ), brackets, [ ], and their more complicated counterparts in mathematics (braces, { }, etc.) are paired off. However, a single paren may be used after numbers or letters — 2) or b), for example — in a list of items. Make sure that the system used is consistent (see Ref. 7, p. 231).

(14) Spacing Between Mathematical Operators, Numerals, and so on

Mark spaces before and after mathematical operators ($+$, $-$, $\times$, $\div$), except when $+$ or $-$ indicate positive or negative numbers ($+7$, -10). Mark spaces between numerals and units of measure (5 mg), and put degree signs close up to temperature symbols, with a space between the number and the sign (20 °C) — unless your journal's house style is to close these up to each other (5mg, 20°C).

When a number has more than four digits, a space should appear after every three digits to the left or right of a decimal point (10 567, 0.159 65; but 1567, 0.1596), in preference to commas.[8] Mark spaces or insert commas where needed, depending on your house style. If columns in tables have a mixture of numerals with four digits and numerals with more than four digits, space all of them after the third digit if the triplet-spacing style is to be followed.

(15) Acknowledgments

Acknowledgments in journal articles may be printed between the text and the references, or as a footnote. If necessary, re-position acknowledgments for the typesetter (cut and paste them, retype them, or photocopy them, if the author has put them in the wrong place). Give instructions about the style of the heading, if there is one, and the type size, if it is to be different from the text type.

(16) Appendix

If there is an appendix, has it been placed where journal practice requires, for example between the text or acknowledgments and the reference list, or after the reference list? Mark the heading style appropriately and indicate the type size, if it is to be different from that of the text. Mark the rest of the appendix in the same way as for the body of the text.

(17) Footnotes/Endnotes

Check that non-bibliographic footnotes (if any) are correctly linked to a reference number, letter, or symbol (call-out signs) in the text, and mark the signs according to journal style. Small superscript signs are the commonest kind of call-out. If symbols are used, use them in the sequence *, †, ‡, §, ||, ¶, doubling them up if more than six symbols are needed (**, ††, and so on). Put the call-outs at the ends of sentences whenever possible. Write "footnote b," or whatever the sign is, in the margin near each sign in the text and circle it. Write "footnote," circled, beside any footnotes typed in the body of the text, and say which size of type should be used.

Some journals abolish footnotes by printing them in the body of the text, between parentheses. Other journals collect the footnotes together and print them as endnotes. Endnotes may be numbered in the same series as numbered bibliographic references and be mixed with them in a list at the end of the article (as is done in the weekly journal *Science*); or there may be two sets of call-out signs (letters and numbers, or symbols and numbers) and two lists at the end. If your journal uses one of these methods, position and mark the footnotes or endnotes appropriately for the typesetter.

References

1. Van Buren R, Buehler MF.   1980   The levels of edit, 2nd ed. Jet Propulsion Laboratory, California Institute of Technology, Pasadena, CA (JPL Publica-

tion 80-1), 1980.

2. ISO/DIS [Draft International Standard] 9115. Bibliographic identification (biblid) of contributions in serials and books. ISO, Geneva.

3. CBE Style Manual Committee.   1983   CBE style manual: a guide for authors, editors, and publishers in the biological sciences, 5th ed. Council of Biology Editors, Bethesda, MD, 1983.

4. Judd K.   1982   Copyediting: a practical guide. Kaufmann, Los Altos, CA, 1982.

5. University of Chicago Press.   1982   The Chicago manual of style, 13th ed. University of Chicago Press, Chicago, 1982.

6. Hart H.   1983   Rules for compositors and readers at the University Press, Oxford, 39th ed. Oxford University Press, Oxford, 1983.

7. Butcher J.   1981   Copy-editing: the Cambridge handbook, 2nd ed. Cambridge University Press, Cambridge, 1981.

8. Symbols Committee of the Royal Society.   1975   Quantities, units, and symbols, 2nd ed. Royal Society, London, 1975.

9. Dodd JS (ed.).   1986   The ACS style guide: a manual for authors and editors. American Chemical Society, Washington, DC, 1986.

10. Windholz M et al (eds.).   1983   The Merck index: an encyclopedia of chemicals, drugs, and biologicals, 10th ed. Merck, Rahway, NJ, 1983.

11. Marler EEJ (compiler).   1985   Pharmacological and chemical synonyms: a collection of names of drugs, pesticides and other compounds drawn from the medical literature of the world, 8th ed. Elsevier, Amsterdam, 1985.

12. Weast RC (ed.).   1982   Handbook of chemistry and physics, 63rd ed. CRC Press, Boca Raton, FL, 1982.

Chapter 6

Technical Editing (Mechanical Style) — References, Tables, Illustrations

References, tables, and illustrations loom large in the copyeditor's working day. Marrying citations in the text to entries in the reference list, styling the list, making sure that tables are comprehensible and figures reproducible, and checking that tables and figures correspond to what is said about them in the text (Checklist 3) can take up as much time as all the rest of the work described in Chapters 3–5.

References

Systems of Referencing

Most scientific journals use one of three systems of citation in the text, with bibliographic details listed at the end of each article. Some journals, however, put bibliographic information in footnotes on each page, with or without a complete list at the end. A few journals use rare systems unrelated to the three main families. The three families described below vary within themselves, as well as between themselves, in punctuation, typography, and the sequence and number of elements required.

The *name-and-year* or *name–date ("Harvard") system* uses authors' names in the text, with the year of publication in parentheses, or the names and year may both be in parentheses: "As Smith & Jones (1985) have pointed out" *or* "As has been pointed out (Smith & Jones 1985)." "And" may be used instead of the ampersand between authors' names, depending on house style.

If the references in an article include two or more papers published

---

Checklist 3   Checklist for mechanical style and format (references, tables, figures)

1. All citations in the text included in the reference list?
   All references in the list cited in the text?
   Names and dates cited in the text agree in spelling and year with those in the reference list?
   Form of citation in the text matches the journal's style?
2. All necessary elements present in the reference list?
   Elements in the required sequence and form?
   Punctuation correct?
   Typographical style marked correctly?
   Journal titles abbreviated correctly (if at all)?
   (Etc.: see Checklist 4)
3. Tables and figures referred to in the text in correct numerical order?
4. Tables and figures show what the text says they show?
   Numbers and other references to table and figure contents match any details cited in the text?
5. Tables have clear titles, column headings, and footnotes, as needed?
   Abbreviations explained?
   All footnote indicators correctly attached to footnotes?
   Numbers in table columns add up correctly (if appropriate)?
6. Layout of tables clear to the reader?
   Different parts of tables clearly marked for the typesetter?
7. Line drawings clearly drawn?
   Photographs of good quality?
   Scale bars included if necessary?
   Lettering, lines, and symbols likely to remain legible if reduction necessary?
8. Legends agree with what the fiigures seem to show?
   Symbols and abbreviations explained?

---

in the same year by the same author or authors, "a," "b," "c," and so on are added to the year: "(Smith 1985a, b, c)."

If there are more than two authors or more than three — depending on house style — all the names may be given in the text the first time the citation appears ("Smith, Braun, and Jones 1999"), with ("Smith *et al.* 1999") or ("Smith et al 1999") without a period after "al," for the second and any other citations. Alternatively "Smith et al. (1999)" or "(Smith *et al.* 1999)" may be used for all citations, including the first. "Et al." or "*et al.*" (for *et alia*) means "and others" and should be used in place of two or more names (Smith, Braun, and Jones), never instead of only one name (Smith and Braun).

The name-and-year system is used in association with an alphabetical reference list in which the year of publication is usually placed immediately after the authors' names. Placing the year anywhere else makes it difficult for readers and copyeditors to find the right reference when there are several references with the same first author.

A second system for references, the *sequential–numeric system*, uses numbers in the text, with or without authors' names ("as Smith[12] reported" or "as already reported[12]"). The numbers usually run sequentially from the first citation to the last, and the references in the list are arranged in numeric order. If the same reference is cited more than once, it often keeps the same number but a few journals assign new numbers to the second and later citations.

The numbers in the text may be small superscript numbers, sometimes in parentheses or brackets, [2], [(2)], [[2]], or with a closing paren, [2)]. Or they may be written on the line, within parentheses or brackets: (2), [2]. In reference lists the numbers are most often written on the line and followed by a period, without parentheses or brackets.

One version of the sequential–numeric system is known as the Vancouver system, discussed on p. 80. This style for references forms part of the "Uniform requirements for manuscripts submitted to biomedical journals" drawn up by the International Committee of Medical Journal Editors[1] and supported by over 300 medical journals.

The third referencing system is a hybrid, usually known as the *alphabetic–numeric system*. The reference list is alphabetically arranged and numbered; the numbers are then inserted in the appropriate places in the text. This system of non-sequential numbering is used much less often than the other two systems.

If the journal uses one of the rarer reference systems not described here, study the instructions to authors or a recent issue of the journal for details of the system.

Most journals require the system laid down in their instructions to be followed to the last comma or colon. Most manuscripts fall short of this ideal — hence the demands that references make on copyediting time. A few permissive (or pragmatic) journals let contributors choose their own reference style, provided that each manuscript is consistent within itself and that enough information is given to allow readers to find the references in a library or retrieve them from a database. Also, references prepared according to the Vancouver style are acceptable to all the journals that support the "Uniform requirements," even though a particular journal may in practice use a variant of this style.

Cross-checking Citations in the Text and Reference List

When you start work on a journal, study its reference system as expounded in the instructions to authors and demonstrated in the pages of the journal. Then deal with the references either during your first reading of a manuscript or as a separate job from editing the text.

First search the text, tables, and figure legends for citations, whether these are names–dates or numbers. As you come across each citation, find the corresponding reference in the list and put a check mark against it — a task that is as essential with numeric systems as it is with names–dates. The check marks, or their absence, will show whether any references are missing from the list or whether any references in the list are redundant — or have perhaps been left out of the text by mistake. Check that authors' names, where these are given in the text, are spelled the same way as in the reference list, and that the dates, where used, are the same in both places.

Unless you work for a permissive journal, correct the form of any citations in the text that don't match the required style.

If the journal requires references to "personal communications" or "unpublished work" to be placed in the text only (see the journal's instructions to authors), remove this kind of entry from the reference list. If the journal frowns on references "submitted for publication," change these to "unpublished work" and restrict them to the text only; if this type of reference is to stay in the reference list, the author must usually be asked to name the journal to which the article has been submitted. References that are described as "in press" may have to be taken on trust, but authors should be asked to fill in the volume and page numbers if these are known at the proof stage.

Styling the Reference List

Next, check the items in Checklist 4 and style the reference list in accordance with the journal's preferred system. Styling reference lists usually leads to various questions on the items in Checklist 4 and on other aspects of referencing. In addition to the material presented below, the journal's instructions to authors, its published reference lists, or the relevant section of the *CBE Style Manual*[2] (p. 50–65) will provide some answers.

Publishers' names and addresses.   When books or parts of books are cited, publishers' names may be given in a shortened form, not the full official name. When it is house style to use the shortened form, delete first names and initials unless these are essential identifiers, and delete "Publishing Company Ltd," "Publishers Inc.," and other such phrases, but keep "Books," "Press," and their equivalents in other languages if these are part of the name (Edward Arnold stays the same because there is at least one other publisher called Arnold; Marcel Dekker becomes Dekker; Pergamon Press stays the same) (see Ref. 3, p. 168–176).

If you or the author happens to know that the original publisher Q of Rville is now part of publisher X in Yville, you might add this informa-

Checklist 4   Basic checklist for styling reference lists

1. References complete?
   a. Names and initials or family name, given name, and initial (or initials) of all authors included (but see 4 below)?
   b. Date of publication included?
   c. Article or chapter titles included, if required?
   d. Journal and book titles included?
   e. First and last page numbers of articles given, if required?
   f. Editors' names, the publisher's name, and the place of publication — or whatever is required — given for books?
2. Sequence of references in alphabetical lists correct?
3. Sequence of elements within each reference correct?
   a. Initials and family names inverted for all authors or just the first author?
   b. Position of date?
   c. Publisher's name before or after the city of publication?
      (And so on)
4. Authors all to be included in each reference, or cut-off point at a certain number of names?
5. Form of journal titles correct?
   a. Titles abbreviated?
   b. If so, abbreviated correctly?
6. Form of references to books correct (edited books, books in series, etc.)?
7. Punctuation correct?
   a. Periods used after initials?
   b. Semicolons used between names?
   c. Period after the year and at the end of each reference?
      (And so on)
8. Typographic style correct?
   a. Initial capital letters used for article titles, journal titles, book titles?
   b. Italics or bold type used for journal or book titles?
   c. Volume numbers changed from roman to arabic, when necessary (use of arabic numbers preferable, even if the cited journal itself uses roman numerals)?

tion in brackets — unless journal policy is against this. "J. & A. Churchill, London [now Churchill Livingstone, Edinburgh]" provides useful information for readers trying to buy a book or get permission to reproduce material from an earlier publication.

   Abbreviation of journal titles.   If your journal prints abbreviated titles for journals in its reference lists, which method of abbreviation is used? Some journals still use the "World List" system based on abbreviations in the three-volume *World List of Scientific Periodicals.*[4] Many more use the "International List" system,[5] which is also observed by the American[6] and

British[7] standards for the abbreviation of titles of periodicals. *Index Medicus,*[8] *Chemical Abstracts Service Source Index,*[9] and *Serial Sources for the BIOSIS Data Base,*[10] among others, follow these standards. If you work in the life sciences, the BIOSIS list[10] is easy to obtain and use, and it shows how nearly all the journal titles likely to be cited in the average biomedical journal are abbreviated. For journals not listed there, or if you work in disciplines for which there are no equivalent lists of abbreviated titles, use one of the standards for title-word abbreviations mentioned above.[4,5,6,7]

Alphabetization of references in the name-and-year system.   Check that the order of references in the list is correct, as well as the order of the different elements making up each entry. Alphabetization is not straightforward when "Smith et al" is used in the text for all references to Smith with two or more co-workers. A practical way to arrange the list, recommended by the Commission of Editors of Biochemical Journals,[11] is to place references to Smith alone first, in their proper alphabetical position among the S's, with the earliest date first. References to Smith with one co-author are given next, arranged secondarily in alphabetical order of the co-authors' names (Smith and Jones 1986, Smith and Thomas 1978, Smith and Whizzkid 1974). Lastly, because the date printed in the text is the reader's only clue to which is which among the "Smith et al" references in the reference list, the remaining "Smith et al" references are arranged secondarily in *chronological* order, earliest date first (Smith, Whizzkid, Brown, and Black *1978*; Smith, Thomas, and Brown *1984*; Smith, Jones, Thomas, Black, and Green *1986*).

If the author has arranged the entries in the list wrongly, use arrows to show the typesetter which references should go where, provided that the new position is on the same page of the manuscript as the old one. If references belong on a different page of the manuscript, write them out in the correct place or cut the list up (make a photocopy first if you have only one copy of the list) and paste it together again in the correct order. But never cut and paste a hard-copy printout if a compuscript has to be corrected on-screen; if you do, the person making the corrections will have a problem finding the parts that have been moved. Instead, use numbers or letters with insert marks (carets) to show which references should be moved where in the printout.

For names with particles such as van, der, and so on, arrange the references under the first letter of the particle, except when the particle is not used in the text or (of course) when the journal has a different rule on this. If "der Schmitt" is cited as "Schmitt" in the text, the entry should go under "Schmitt A der," *not* under "der Schmitt A": the style of the citation in the text should match the style in the reference list. For hyphenated family names, arrange entries in alphabetical order under the initial letter of the first part of the name. For a double name with no hyphen, alphabetize under the second name (J. Smith Jones becomes "Jones JS" or "Jones J Smith").

When authors add "Jr" or "III" to their names, and if names and initials in the reference list are inverted, as they often are, put Jr or III after the initials ("Jones JS Jr"); or omit these dynastic indicators if journal practice allows you to do so.

Numbers and years in sequential–numeric and alphabetic–numeric systems. In numbered reference lists the numbers are usually written on the line, followed by a period, but parentheses or brackets are sometimes used. In the sequential–numeric system the year of publication usually goes either before the journal volume number or after the page numbers. In references to books the year may go at the end of the reference. In the alphabetic–numeric system the year may go in the same position as in the sequential–numeric system or immediately after the authors' names, as in the name-and-year system. (In reference lists in this book, the year appears in both positions.)

Dealing with Incomplete References

If you have easy access to a suitable library, it may be quicker to complete or correct imperfect references yourself than ask the author for the information. The *Science Citation Index, Index Medicus*, and other reference sources, with details gleaned from the journals themselves or from an on-line database (if your journal is rich enough to pay the search charges), can all help. Use your common sense to decide when you have spent enough time hunting for elusive information in obscure journals; your employers will be getting poor value for their money if you spend all day tracking down half-a-dozen page numbers for articles said to be in the *Ruritanian Geological Society Journal* for 1890.

Verifying References

Even though a perfect reference list is rare, few journals go to the trouble of verifying from the original sources that authors' references are correct. If this task nevertheless falls to your lot, the time you spend in the library will have the editor's blessing. More often the practice is either to spot-check a few references or rely totally on what the author supplies, making a check only when queries arise, such as when elements of a reference are missing or don't match the elements given in other references or citations.

Writing Queries About References for the Author

Write queries in the margins of both the text and the reference list (or write them on flags or list them separately). Include cross-references from the manuscript page numbers to queries in the list, and vice versa ("Or 1980, as in refs.?" on p. 8, with "Or 1985, as on p. 8?" opposite a 1980 reference in the list, for example). If authors are to receive proofs later, your queries should include requests for the volume and page numbers, or other information about "in press" references, to be added if the articles have been published between manuscript and proof stages.

Towards Uniformity in Reference Styles

If you aren't already appalled by the variety of reference styles in scientific publications or in the manuscripts you have to cope with, you probably soon will be. Two groups of editors have tried to produce some order from the chaos. In 1978 the European Life Science Editors' Association (now the European Association of Science Editors) recommended a reference list style that can be used with either numbers or names-and-years in the text. This system, known as the ELSE–Ciba Foundation or ECF style,[12] is basically as follows:

Adam A E, Eve B C 1978 Preventing rot in Mediterranean apple
trees. Journal of Tree Studies [*or* J Tree Stud] 90:21-29

In 1979 the International Committee of Medical Journal Editors (ICMJE) outlined the "Vancouver style," which requires numbers to be used in the text and date of publication to be placed before the volume and page numbers in the reference list. This system[1] is almost as economical of keystrokes as the ECF system:

Soter NA, Wasserman SI, Austen KF. Cold urticaria: release into
the circulation of histamine and eosinophil chemotactic factor of
anaphylaxis during cold challenge. N Engl J Med 1976; 294:687-90.

The minimal punctuation and simple typography of the ECF and ICMJE styles obviously save typing and typesetting time. Both styles follow the main recommendations of the American national standard for bibliographic references.[13] Neither that standard nor any other national or international standard, however, has yet dealt satisfactorily with all the problems of bibliographic references in scientific publications. In allowing for either numbers or names and years to be used, the ECF recommendations face up to the main obstacle to uniformity in referencing: that is, that the same authors submit papers to different journals and these journals may demand either of the two main reference systems. Both styles in fact have a valid role in different publications. They may both, however, need minor changes in punctuation in the era of electronic publishing.

Chaotic reference systems may not be your first interest when you start work as a copyeditor, but it won't be long before you realize why references take up a disproportionate amount of your time. At some stage in your copyediting career you may be able to influence a journal's choice of reference style and encourage a move to an economical system of the ECF[12] or ICMJE[1] kind. If you are an authors' editor you can advise your authors to keep full bibliographic records in one of these forms (including authors' first names in full when these are given on the original article). Both styles can be easily adapted, when necessary, to more heavily punctuated or typographically varied versions, with or without the aid of computer programs to manipulate the details. Computer programs for personal bibliographic databases are a boon, of course, but bibliographic details have to be recorded in full before they can be reshaped electronically or manually to the required style.

Tables

The editor and referees will have assessed the quality of the tables in manuscripts and decided whether they are appropriate and essential. Your role is to make sure that tables have been referred to in the text in the correct numerical order, calculate whether they will fit the column or page, and check the numbering, titles, footnotes, headnotes, abbreviations, and column headings and contents. Then you must indicate how tables should be set, clarify which part is which for the typesetter, and ensure that table contents agree with statements in the text.

The Parts of Tables

First, the terminology. As well as titles, tables may have headnotes or footnotes, or both, and some or all of the following: column headings or subheadings, straddle (or spanner) rules, side (or stub) headings, and the field. Straddle or spanner rules span several columns, each with its own column heading or subheading. Side or stub headings are the entries in the first column (or stub) in many tables; stub headings may have subheadings too. The field is the area under column headings where the findings are entered (see Fig. 9 and Ref. 2, p. 75–76).

Position in Text

Write "Table 1" and so on in the margins of the text beside the first reference to each formal (numbered) table and circle this message. The typesetter and (or) the person who pastes up or arranges the final proof for

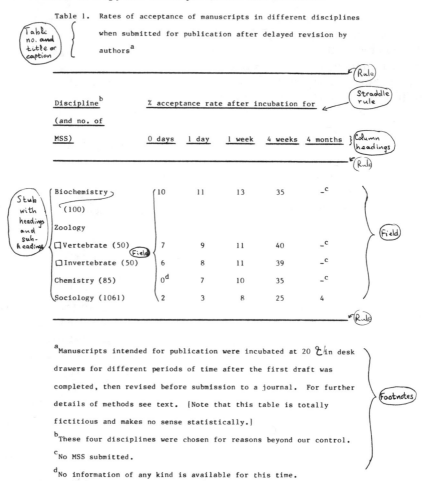

Figure 9   The parts of a table.

printing will then know approximately where to place or leave space for the tables in the printed text — "approximately" because the constraints of the column or page size may mean that tables can't be printed immediately after they are first referred to (the preferred position for a table is usually at the top or bottom of a page, as close as possible to the first reference to it in the text). This notation in the margin also shows whether all the tables have been referred to and whether they have been numbered consecutively, in the order of their first mention in the text.

Size of Tables

If you want to be sure that a table will fit the space available, assess its width in characters (letters, numbers, spaces), adding two characters for the space between each column. Compare the number of characters with the number in an average table printed in the journal (keep a note of this second number to refer to when necessary).

If a manuscript table seems too wide after you have calculated its size in this way, examine the possibilities for reducing the width. Remove any unnecessary words from the stub entries but otherwise leave the typesetter to decide where the runover (or turnover) to a second line should start, if a second line is needed. Make all the column headings as short as possible, using footnotes to expand them if necessary — but avoid abbreviations other than recognized ones or those used (and explained) in the text and in the footnotes to the table. Keep the columns themselves narrow by placing units of measure in the column headings or stub headings instead of allowing them to be repeated beside each entry (see "Column Headings and Contents" below).

If the table is still too wide after these operations, will switching the stub headings with the column headings reduce the width while still getting the author's message over clearly (consult the editor or the author, or both)? If a 90-degree turn of this kind is not feasible or still fails to reduce the width enough, the table will have to be set sideways on the page; ask the typesetter to turn it or set it "broadside" (United States) or "landscaped" (United Kingdom), or in a smaller size of type than usual (ask for 8-point type, for example, if 10-point is the regular size for tables).

If very long tables take up more than one column or page when printed, ask the typesetter to start the second column or page with "Table 5 continued" (or whatever style the journal uses) and ask for the column headings to be repeated. If the headings have been repeated on each manuscript page of a long table, circle all of them except those on the first page.

Numbering, Titles, Credit Lines

Check that the manuscript reference number appears on tables in the top right-hand corner, circled to show that it is not to be printed. A short title should follow the table number. Check that in a series of similar tables the titles are parallel in grammatical form. Remove from titles phrases such as "a study of" and any information also contained in the column headings. If a table has been borrowed from or is based on an earlier publication, add a reference at the end of the title ("from Smith 1980" or "based on Green 1985") or in a footnote. The copyright-holder may ask for a full reference to be included in the credit line; make sure that the reference is included in the reference list too.

Headnotes/Footnotes

Headnotes or footnotes to tables may be used to give quite lengthy details of experimental methods. Edit these as carefully as you edit the rest of the text. Other kinds of explanatory notes are also usually printed below the main body of tables. If typed tables do not follow the journal's style for notes, mark the typescript to show that a headnote should be set as a footnote, or a footnote as a headnote, as necessary.

Depending on house style, footnotes describing methods may or may not be linked to positions in the table by a superscript letter, number, or symbol. Check that other kinds of footnotes to tables *are* so linked and that each superscript character appearing in the title or in the column headings or body of the table has a corresponding footnote. Are the characters the kind the journal prefers? If they are symbols, are they a sort the typesetter has available — usually *, †, ‡, §, ||, ¶, then **, ††, and so on — and are they used, as they should be, in the sequence printed here? In the field or body of the table place these superscript characters to the right of the words, numbers, or abbreviations to which they refer, and make sure that they appear in order first from left to right, starting from the top of the table, and then from row to row, as in Fig. 9.

Column Headings and Contents

Column headings within tables need to be very brief (see "Size of Tables" above). They must also be clear, unambiguous (see Chapter 3), and parallel in style. When total values for columns or rows are given in a table, check the arithmetic. If there are mistakes, correct them, and correct any mention of these numbers in the text. Draw the author's attention to the changes you make: "2140 OK? see Table 3" on page 5 and "2140 OK? see p. 5" on the table.

If several different units of measure are used in the same column or row, see whether they are suitable for conversion to the same unit. This allows one unit to be included in the column heading instead of a unit being printed beside each entry in the field. If the units are the same but have nevertheless been typed after each entry in the field, delete them and put the unit in the appropriate column heading.

If the author has used ditto signs anywhere in a table, delete them and write the words or numbers in full. If minus signs are used in a column that also contains a range (e.g., 20–30), substitute "to" for the dashes ("20 to 30").

Check whether a blank in a column indicates — as it should — that the column heading is not relevant to that particular stub heading. ND (for "no data") should be used (and explained in a footnote) when a result has been looked for but not obtained. If dashes are used for this purpose, explain the dashes in a footnote (check with the author that the explanation is correct).

Abbreviations

Explain all the abbreviations in tables if the author hasn't already done so (footnote symbols are often used with abbreviations but a simple list without linking symbols is allowed too). Ask the author to expand any abbreviations you can't decode. Strictly, all abbreviations should be explained in every table, even if the same ones appear in several tables or have also been explained in the text: many readers look at the tables before reading the text, and some look at the last table first. But follow your journal's practice; saving space may be more important than providing for all possible reading habits.

For the rest of what has to be done to tables before they are typeset, see Chapter 7.

Illustrations (Figures)

As with tables, the editor and referees will have decided whether illustrations (figures) are acceptable in quality and quantity, but the copyeditor may check the quality too (see Chapter 3). Again as with tables, if a figure seems to give little or no useful information, consult the editor or author about removing it. Check that a marginal note appears beside the first mention of each figure to indicate approximately where it should go in the printed article, and make sure that the figure numbers run consecutively.

Then check that figures are indeed numbered and that the author's name and the manuscript reference number are written on them. Write "Top" in the appropriate place, even if it doesn't seem to be essential. For photographs write this information either on the front of the figure if you can do this outside the part that will be reproduced, or write it very lightly in pencil on the back. Or, as marks show through easily and may show up in the printed version, write the information on an adhesive label and fix this to the back of the photograph. Because marks show up so easily, avoid using paperclips on photographs.

Redrawing Figures

Line drawings (graphs, for example) may be submitted either as original drawings or as photographs (photoprints) of originals. Some journals have line drawings redrawn to give their pages a unified appearance, or they may ask for lettering on axes, curves, and so on to be supplied on an overlay so that the figures can be lettered in journal style. If your journal goes in for redrawing, edit the lettering and make other essential changes on photocopies of the originals or on the overlays before you send the figures to the artist. Send photocopies of redrawn figures, not the drawings themselves, to the author for checking, if journal practice allows this.

Rather than have figures redrawn or remade, many journals accept them in the form supplied by authors, or the authors may be asked to correct them, if necessary. If authors are responsible for making new figures, check whether the revised version has clear lines, lettering, and symbols (see "Lettering and Symbols" below) and that photographs are of a good standard (good contrast, sharply focused).

Sizing Figures

If your job includes marking the reductions for figures, use a reduction wheel to calculate the likely percentage reduction (x), or calculate $x\%$ by measuring the width of a printed column or page (y) in the journal and the width of the original figure (z); $(y \times 100)/z$ will give you x. To fit a 24 pica column (9 cm, or 3.5 in) (see Chapter 7), a figure 15 cm wide (about 6 in) would have to be reduced to 60% of its original size: $x = (9 \times 100)/15$. Write "$\times$ 60%" or "shoot at 60%" lightly on the back of the figure (on an adhesive label, if necessary). Check that "60%" doesn't mean a reduction *of* 60% — i.e., to 6 cm instead of 9 cm — in your printing-house. If a figure is to remain the same size, write "S/S" on it. In addition, you may need to make a list of the figures and the reductions they need, for the production editor. If only part of a figure is to be reproduced, put pencil marks (crop marks) on the back or edges to show the area that is to be used, or mark this area on an overlay, or mask the unwanted parts by taping paper over them.

Lettering and Symbols

Are letters and symbols open enough not to blur and are the lines or curves thick enough not to vanish if the figure is reduced? If an original line drawing or a photoprint (a photograph or photocopy of a line drawing) is likely to be reduced by a third, the lettering should be about 2.5 mm high on the original or photoprint, to produce characters of not less than the recommended minimum height of 1.5 mm (4 points: see Chapter 7) in the printed version. The final size of lettering on photographs should be not less than 2.5 mm (7 points) (Ref 2, p. 73).

If lettering or symbols have been gummed to figures, make sure that they are firmly fixed.

If drawings or photoprints arrive with titles lettered on them (titles that are or should be in the legends), mask the title with an opaque overlay or attach a note saying that it is not to be included in the printed version (unless, of course, it is journal style to keep titles on the illustrations themselves). If there is no need to return the original figures to the author, white out unwanted titles or cut them off.

For graphs, see that all abscissas (horizontal or x axes) and ordinates (vertical or y axes) have been identified in suitably brief wording and clear lettering. Curves in graphs should be identified by different kinds of lines or by different symbols for experimental points, though they do not need to be differentiated in both ways in the same graph. If the symbols appear in the legends, has the author used symbols that most typesetters have available, such as $\bigcirc$, $\triangle$, $\square$, $\bullet$, $\blacktriangle$, $\blacksquare$? If an exotic symbol is included, is it explained in a key in the body of the figure? If not, or if the symbol has to be used in the text, consider asking the author to substitute a conventional symbol (or consider asking the editor about this). If the journal redraws line drawings, tell the artist which symbol to substitute, and make the necessary changes in the legend and text.

Scale Bars/Magnifications

Photomicrographs or other photographs where scale is important should preferably have scale bars drawn on them, as these will still indicate the correct relative size after the figures have been reduced for printing. If a magnification number (e.g., $\times$ 1200, or 1200 $\times$) appears in the legend instead of a scale bar, you may need to change the number or mark it to be changed when the final reduction of the printed version is known (which may not be until the proof stage). Some publishers prefer to give both the original magnification and the photographic reduction ("$\times$ 1200, photographic reduction 40%"). If neither scale bars nor magnifications are given on figures that need them, ask the author to supply the information at the revision stage; if the author doesn't see the copyedited manuscript until the proof stage, write "$\times$ 0000" in the legend (using a suitable number of zeros, if you can), with a note asking the author to fill in the correct number.

Legends

Each figure must have a legend consisting of a short title and any further description that is needed, including explanations of abbreviations and symbols. The legends should be listed on a separate sheet or sheets of paper, not on the figures themselves (unless the journal requires this). Examine the legends to make sure they match the figures they ostensibly describe; then edit them into a consistent form and style, according to journal practice. Place descriptive titles after "Figure" or "Fig." and the figure number (but you may need to remove descriptions such as "Photograph of": see p. 57), and place other explanations or comments after the title. Place magnification numbers (e.g., $\times$ 1200: see above) after the explanations of abbreviations and symbols, or wherever house style requires them to go. Check that all abbreviations and symbols, apart from mathematical or other recognized symbols, are explained in the legends (and see p. 87 on unconventional symbols). Query anything that hasn't been explained, but first make sure you haven't overlooked an explanation in the text that could or should be added to a legend. The reader may forget the explanation too, or — as for tables (above) — may look at the figures before reading the text.

If a figure has been reproduced from a previous publication, add the acknowledgment or credit line at the end of the legend, or wherever the journal usually puts it, if the author hasn't already seen to this. Some copyright-holders ask for full reference details to be included in legends when figures are reproduced from published work. Make sure that the reference is also included in the reference list.

Wrapping It All Up

If your work on a manuscript ends at this stage, assemble the parts of the article in the order preferred by typesetters and production editors and, if necessary, prepare manuscripts for mailing (see p. 28). If your responsibilities extend further into the area of production and mark-up, Chapter 7 describes the next stage.

References

1. International Committee of Medical Journal Editors.   1982   Uniform requirements for manuscripts submitted to biomedical journals, rev. British Medical Journal 1982; 284:1766–1770/Annals of Internal Medicine 1982; 96:766–770 [original publication: International Steering Committee of Medical Editors. 1979 Uniform requirements for manuscripts submitted to biomedical journals. British Medical Journal 1979; 1:532–535] (and elsewhere).
2. CBE Style Manual Committee.   1983   CBE style manual: a guide for authors, editors, and publishers in the biological sciences, 5th ed. Council of Biology Editors, Bethesda, MD, 1983.
3. Judd K.   1982   Copyediting: a practical guide. Kaufmann, Los Altos, CA, 1982.
4. Brown P, Stratton GB. (eds.).   1963–1965   World list of scientific periodicals published in the years 1900–1960. Butterworth, London, 1963–1965.
5. ISO 833–1974. Documentation — international list of periodical title word abbreviations. International Organization for Standardization, Geneva, 1974.
6. ANSI Z39.5–1974. Abbreviation of titles of periodicals, rev. American National Standards Institute, New York, 1974.
7. BS 4148–1970, 1975. Abbreviation of titles of periodicals. Part 1: Principles, 1970; Part 2: Word-abbreviation list, 1975. British Standards Institution, London, 1970, 1975.

8. Index Medicus. [Issued monthly, consolidated annually] National Library of Medicine, Bethesda, MD.
9. Chemical Abstracts Service Source Index.
10. BioSciences Information Service. [Annual] Serial sources for the BIOSIS Data Base. BioSciences Information Service, Philadelphia.
11. IUB-CEBJ [International Union of Biochemistry-Commission of Editors of Biochemical Journals]. 1973 The citation of bibliographic references in biochemical journals. Recommendations (1971). Biochemical Journal 1973; 135:1-3 [and elsewhere].
12. [ELSE]-Ciba Foundation Workshop. 1978 References in scientific publications: suggestions from a Ciba Foundation Workshop, London, 25 November 1977. Earth & Life Science Editing 1978; No. 7: p. 18-21 [and elsewhere].
13. ANSI Z39.29-1977. Bibliographic references. American National Standards Institute, New York, 1977.

Chapter 7

Marking Up and Coding
Manuscripts and Compuscripts

When mechanical style and the aspects of format discussed in Chapters 5 and 6 have been dealt with, more instructions will have to be given about visual style — unless the journal is produced directly from author-prepared camera-ready copy (see Chapter 2). That is, you must identify the constituent parts of a manuscript or compuscript on its way into print; mark up or code those parts to show which type faces, sizes, and weights are to be used; and mark any sections or characters needing special typographical treatment

Typography

Typographical mark-up is sometimes done by the typesetter's staff, sometimes by copyeditors. In either case some background information on typography is essential (see Ref. 1, p. 93–95; Ref. 2, p. 561–576, p. 52–56, p. 72–75; Ref. 3, p. 177–195).

The terminology of type and typesetting (or composing, as it is also called) goes back to the fifteenth century, to the time of Gutenberg and Caxton, when individual letters and other characters were cast in a lead alloy (hence "hot metal") and stored in flat cases or trays with divisions for the different characters. Capital letters were usually kept in the upper case of a pair of cases which rested on a frame in front of the compositor, small letters being kept in the lower case (see Ref. 4, p. 49–65; Ref. 5, p. 29).

Type faces come in many different designs, with names such as Times Roman, Baskerville, Bembo, Garamond, and many others. You are not expected to recognize all type faces instantly, but several of these will soon become familiar.

A type font, or fount, is a set of alphabets in one type size and face: capitals and lowercase characters in roman (upright type) and italics (sloping type), and small capitals, making five alphabets — or seven if bold capitals and lowercase are included. Some fonts, for example 18-point, 14-point, and 12-point Univers and Helvetica, are popular for titles, headings, and other places where display type is needed.

Univers and Helvetica belong to the type style known as sans serif; that is, they have no serifs or cross-lines at the ends of the main strokes of the characters.

A point is a measure used for the height of a type face, from the top of a capital letter or a letter with an ascender (such as "d") to the bottom of a letter with a descender (such as "p"). A point in the United States and the United Kingdom measures 0.35 mm or about 0.014 in (about $\frac{1}{72}$ in); a point in the Didot system used in Continental Europe measures 0.376 mm (0.015 in). Points are also used to describe the thickness of printed lines (rules) and the spacing between lines, words, or characters.

The measure used to specify the width of type on a printed page or column is the pica, which measures 12 points (4.2 mm, 0.165 in). An instruction to set columns about 100 mm wide (4 in), for example, would read "× 24 picas."

The area of the type page (the part that has printing on it, including the page number and the headline or footline) may be described in terms of its "emage" or number of ems (width × height) in the type size being used for the text. An em or "em quad" is as wide as a type face is high — in 10-point type an em is 10 points wide, in 18-point type it is 18 points. An en is traditionally half an em in the same point size, but in computer typesetting the exact size varies a little, depending on the particular system being used. Spaces finer than the en may be needed: thick, thin, and hair spaces measure from one-third to one-fifth of an em. Lateral spacing may also be measured in either points or "units": an em quad is 18 units wide, whatever the type size. Typesetters sometimes use measurements in millimeters or units instead of the point and the pica.

The space occupied on the printed line by an alphabet in a particular type face depends on the width, not the height, of characters. Different type faces of the same point size may have different character counts to the centimeter or inch, which is something designers have to remember when choosing type faces for journals or books, especially when space is scarce.

Journals and books are usually printed in 9-point, 10-point, or 11-point type, sometimes with the lines set solid (no extra space between them) but more often with a 1-point or 2-point space, known as leading, between them.

The appropriate instruction for setting text might be "10 on 12 Bembo" or "10/12 Times Roman," which in traditional terminology means 10-point type cast in a lead alloy on a 12-point body, or 10-point type with 2-point bars of metal between the lines to produce white space on the printed page. The text type you are reading is 10-point Times Roman with 2-point leading. But with present-day methods of computer typesetting there is no casting, and type sizes vary slightly from one typesetting system to another.

Most books and journals are printed with the type lines justified (all set to the same measure, with the space between words automatically adjusted by the typesetting system). "Ragged right" setting, where the right margin is uneven, is sometimes used and is said to produce a small saving in space and a slight increase in legibility.

Other terms you may hear are "paste-up" and "stripping-in." In a paste-up, proofs are arranged and pasted in their expected final position, with space left for illustrations if necessary. When the layout has been approved and the illustrations are ready, a good quality proof (a "repro") is pasted up and photographed; the film negative is then used to make plates for printing by photolithography. Stripping-in is the process of combining two pieces of film negative or paper, often for the purpose of making corrections or changes in one or more lines.

You will meet these and many other terms concerned with typesetting and printing before you have been copyediting for long. Typesetting methods and systems, old and new, are described more fully in the *Chicago Manual of Style*[2] (p. 586–591, 600–606, 628–636) and by Jennett[4] and Lee.[5] You will probably find it helpful to visit a typesetting or typesetting/printing house after you have had a few months' experience of copyediting. By then you'll have plenty of questions, you'll be able to see how the marks you make on manuscripts and proofs affect the work of typesetters and printers, and you'll appreciate the difference your work and theirs has made to the appearance of the printed page. (You may be able to see all the processes in producing journals or books under one roof, but typesetting, printing, and binding are often done by different firms in different places.)

Manuscript Mark-up

Before you start marking up a manuscript for typesetting, delete any superfluous marks, comments, or instructions written in the margins or anywhere else. If you don't do this, the typesetter has to spend valuable time looking at these jottings to discover their relevance, or lack of it, to typesetting.

You should also study the design specifications ("specs") for the publication. For a journal the specs will follow a standard pattern, though the design may be reviewed from time to time. Books in a series often follow a standard design too, but for other books keep the specs beside you when you do the mark-up, as well as studying them carefully before you start.

Marking Up Text

If the typesetter holds a standard set of specs for a book or journal, give your instructions about type faces and sizes by keying the various parts of each manuscript to the specs. Specs include the trimmed size of pages, the width or depth of margins, and the type area. They give the type faces and sizes for text and for chapter openings, running heads and folio numbers, headings and subheadings, footnotes, references, tables, table titles and footnotes, and legends for figures. They also provide instructions about the placement of figures, equations, or formulas, and about other matters affecting design, according to the requirements of particular publications.

The specs might, for example, require you to write A, B, and C (circled) in the margin beside headings in the text, indicating their rank and style for the typesetter (see Fig. 10). In some cases, you may have to write "T" beside all the parts that are straightforward text. In fact, simple mark-up of this kind is often done as part of the copy-marking procedures described in Chapter 2. You'll need a list of the keys or codes usually used for the journal, or you may be asked to produce your own list.

Each key identifying a heading refers to a particular type face, size, and weight (bold, italic, or roman). An "A" heading might be translated as "cap. & l.c. bold 12-pt Times Roman, 6-pt leading above, 4-pt leading below." The typesetter's translation table shows what the keys or codes mean for each set of specs.

If the typesetter has no specs for the material you are preparing, state the width of the columns or pages (e.g., 24 picas) on the transmittal sheet (see p. 15) and on the first page of each article or chapter. If the typesetter has an automatic page make-up system, in which pages are laid out on a screen instead of being manually pasted up (see above), state how many lines of text should appear in a full page or column. Mark the type faces, type sizes, and leading (Helvetica Bold 18 pt, Times Roman 10/12 pt, or whatever) wherever these change for different parts of the manuscript.

Point out any special symbols or alphabets by writing the names of the characters in the margin (e.g., "l.c. Greek alpha," circled). A separate list of all the special characters ("sorts") appearing in the copy may be needed too.

The easiest way to see how to mark up a manuscript is to look at the "dead copy" returned by the typesetter after a recent journal issue or book was printed. For journals this will show you which type faces and sizes are

Fibrinogen has been modified proteolytically by a crude bacterial enzyme to produce highly ordered microcrystalline and crystalline forms. Preparative methods are described here, the electron microscope images are surveyed, and preliminary chemical and crystallographic data are presented. The modified fibrinogen is largely intact and highly clottable. Many of the modified fibrinogen aggregates have a 45 nm axial repeat; a number of ordered forms, including fibrin, can be derived from one or two basic arrays. It is inferred that fibrinogen is 45 nm long and non-polar, i.e. that there is a twofold axis perpendicular to the long axis of the molecule. Filaments of fibrinogen bonded end-to-end may be a feature of all these arrays.
Proteolysis, Pseudomonas, microcrystal, fibrin, clottability, electron microscopy, X-ray crystallography.

Fibrinogen is the soluble precursor of the fibrin clot. In vertebrates, the fibrinogen molecule has a relative molecular mass (M_r) of about 340 000 and consists of three pairs of polypeptide chains – Aα, Bβ, and γ, crosslinked by 28 disulfide bridges (Henschen 1974).

(Greek alpha, beta, gamma)

. . .

A   RESULTS

B   Ordered forms of modified fibrinogen

Historically, the ordered aggregate of fibrinogen studies by electron microscopy has been the relatively insoluble fibrin . . .

C   □   Low ionic strength forms: microcrystals. Fibers of native bovine fibrinogen formed at low ionic strength (for example, 0.005 M-KSCN or KCl, 0.01 M-morpholinoethane sulfonic acid, ph 6.2) show little axial order by electron microscopy, although often . . .

Figure 10   Sample of a section of text marked up for typesetting, including abstract, keywords, and three orders of headings. Spacing of headings and other instructions would be covered by the specifications. Compare this with Fig. 13.

used for article titles, authors' names and addresses, headings and subheadings in the text, the text itself, and for tables (see below) and all the other possible parts of the manuscript (see Fig. 10). For books, follow the specs closely.

Table 1. Rates of acceptance of manuscripts in different disciplines when submitted for publication after delayed revision by authors[a]

Discipline[b] (and no. of MSS)	% acceptance rate after incubation for				
	0 days	1 day	1 week	4 weeks	4 months
Biochemistry (100)	10	11	13	35	—[c]
Zoology					
Vertebrate (50)	7	9	11	40	—[c]
Invertebrate (50)	6	8	11	39	—[c]
Chemistry (85)	0[d]	7	10	35	—[c]
Sociology (1061)	2	3	8	25	4

[a]Manuscripts intended for publication were incubated at 20°C in desk drawers for different periods of time after the first draft was completed, then revised before submission to a journal. For further details of methods see text. [Note that this table is totally fictitious and makes no sense statistically.]
[b]These four disciplines were chosen for reasons beyond our control.
[c]No MSS submitted.
[d]No information of any kind is available for this time.

Figure 11   Figure 9 marked up for typesetting.

Marking Up Tables (Fig. 11)

Most tables have horizontal rules at the top and bottom, with a third rule between the column headings and the field or body of the table (see Chapter 6, "Parts of Tables"). The three full-width rules and any straddle or spanner rules showing which headings belong to which columns are the only essential rules in most tables — vertical rules are usually deleted in favor of spacing between columns. Delete any unwanted lines that have been drawn or typed in the manuscript, including lines that box in the whole table (unless, of course, it is journal style to use boxed-in tables).

Mark the letters or words that should be capitalized or italicized in the column headings of tables. Show whether the headings are to be ranged left or centered over the columns they describe. If headings or entries in the first column (stub headings) have been typed ambiguously, mark them to show what their final position should be. Mark any subheadings in the stub for indention by one em or en, or whatever the standard indention may be, and do the same for turnovers or runovers (entries that run over to more than one line) in the stub or other columns — the typed lines may not match the printed lines, but the typesetter will understand what is needed.

Mark entries in the field of a table to show how they line up horizontally with the relevant stub headings; if a stub heading has turnovers, align the entries in the field with the last turnover line of the heading. Show the typesetter whether the entries in the field are to be ranged left or centered in the columns, and whether they are to be aligned on real or imaginary decimal points, or aligned first on $\pm$ signs and then on decimal points. If necessary (see Chapter 5), put spaces between every three digits in either direction from the decimal point but do not space four-digit numbers in this way unless other numbers in the same column have more than four digits.

Correcting and Coding Compuscripts

Even when authors submit their articles on floppy disks or magnetic tapes, you will probably do most of your copyediting on a hard-copy printout. Eventually, however, your changes and the necessary formatting codes have to be entered on the disk or tape. Who should do this work — the author, the copyeditor, or the typesetter?

Journal practice on who keys in corrections, changes, and formatting codes varies according to the hardware and software available and whether interfacing problems between the hardware and software used by authors, the editorial office, and the typesetter have been overcome. If authors check their copyedited printouts before typesetting begins, they may be asked to correct the disks or tapes at the same time and to insert at least the simpler codes. In this case, either add the codes to the copyedited printout and ask the author to see that these are included when the changes and corrections are made, or send the author a list of codes and instructions on entering them when you return the copyedited printout. The remaining codes will

#SS	start (title)	#KK	captions for tables
#NN	names of authors	#LL	legends for figures
#II	address of Institute etc.	#RR	reference section heading
#GG	title page credit line	#RE	each new reference
#XX	abstract	#ZZ	acknowledgments
#TT	text	#FF	footnotes (text)
#AA	A headings	#EE	end
#BB	B headings		

En dash: 2 hyphens
Em dash: 3 hyphens
Symbols: separate list [# with two-digit number]
@:   paragraph indent

Figure 12   Sample set of generic codes for typesetting (*ad hoc* system arranged with typesetter).

be added later by the typesetter. Alternatively — hardware and software permitting — it may be your job to make the corrections and do all the coding, or to check that someone else has done so.

Coding compuscripts, like keying manuscripts, consists of identifying the different parts (elements) and keying in the appropriate codes, as arranged with the typesetter. Different typesetters use different sets of codes, but many coding systems are mnemonic and can be listed on one or two pages (Fig. 12). If the word processor's commands for italic and bold, subscript and superscript, and accented and Greek letters are recognized by the typesetting system, or by the interfacing system between the author's or editor's word processor and the typesetter's terminal, the job of coding can be very simple (see sample page, Fig. 13).

If you are asked to use the generalized system of coding known as the Standard Generalized Markup Language, your job will not be so easy (see Fig. 14). SGML is described as a device-independent method for coding or tagging the structural elements of compuscripts, to allow some or all of the material to be processed for different uses in different typographical forms.[6] An admirable concept — but one that seems destined to make life difficult for whoever has to enter the codes and check that they are correct. Straight text is easy enough to tag and even to read when printed out. The trouble comes with tables and reference lists, which need so much tagging that they are unreadable in printouts.[6,7] By the time this book appears the whole process may have been automated and the problems removed or

#SSCrystalline states of a modified fibrinogen

#NNNancy M. Lavoisier and Carolyn Culpepper

#IIDepartment of Basic Medical Sciences, University of North Schuyler,

1050 Deep Drain Street, North Schuyler, PA 19100-3000

#GG Lavoisier NM, Culpepper C 1987 Crystalline states of a modified
fibrinogen. Quarterly Journal of Fibrogenesis Studies 9:000-000

#XXFibrinogen has been modified proteolytically by a crude bacterial
enzyme to produce highly ordered microcrystalline and crystalline
forms. Preparative methods are described here, the electron
microscope images are surveyed, and preliminary chemical and
crystallographic data are presented. The modified fibrinogen is
largely intact and highly clottable. Many of the modified
fibrinogen aggregates have a 45 nm axial repeat; a number of
ordered forms, including fibrin, can be derived from one or two
basic arrays. It is inferred that fibrinogen is 45 nm long and
non-polar, i.e. that there is a twofold axis perpendicular to the
long axis of the molecule. Filaments of fibrinogen bonded end-to-
end may be a feature of all these arrays.
Proteolysis, Pseudomonas, microcrystal, fibrin, clottability,
electron microscopy, X-ray crystallography.

#TTFibrinogen is the soluble precursor of the fibrin clot. In

vertebrates, the fibrinogen molecule has a relative molecular mass

#60
(M_r) of about 340,000 and consists of three pairs of polypeptide

chains Aα , Bβ , and γ , crosslinked by 28 disulfide bridges

(Henschen 1974).

. . .

#AARESULTS

#BB Ordered forms of modified fibrinogen

#TT Historically, the ordered aggregate of fibrinogen studies by

electron microscopy has been the relatively insoluble fibrin . . .

#CC Low ionic strength forms: microcrystals. Fibers of native bovine
#60
fibrinogen formed at low ionic strength (for example, 0.005 M-KSCN or
#60
KCl, 0.01 M-morpholinoethane sulfonic acid, ph 6.2) show little axial

order by electron microscopy, although often . . . #EE

Figure 13 Text commands (tags, codes) inserted on an edited compuscript printout, using the system shown in Fig. 12. The #60 code ensures that the sequences
of numerals on either side of the code are not split at the ends of lines and that
a (thin) space is inserted between them.

```
<BR><RTID=B10.1> <A>Smith, A. <TI>Spending money too fast.<PCY> London:
<PNM>Economic Publishers,<PDT>1989,<PGN>101—10.</RT></BR>
```

Figure 14   Reference in a single-reference bibliography marked up in SGML (Standard Generalized Markup Language).

minimized — but until that utopian state of affairs has been achieved, the job of entering and checking this kind of coding in scientific material should be one for typesetters, not authors or copyeditors.

References

1. CBE Style Manual Committee.   1983   CBE style manual: a guide for authors, editors, and publishers in the biological sciences, 5th ed. Council of Biology Editors, Bethesda, MD, 1983.
2. University of Chicago Press.   1982   The Chicago manual of style, 13th ed. University of Chicago Press, Chicago, 1982.
3. Judd K.   1982   Copyediting: a practical guide. Kaufmann, Los Altos, CA, 1982.
4. Jennett S.   1973   The making of books, 5th ed. Faber & Faber, London, 1973.
5. Lee M.   1979   Bookmaking: the illustrated guide to design/production/editing, 2nd ed. Bowker, New York, 1979.
6. Association of American Publishers.   1985   Author's guide to electronic manuscript preparation and generic tagging. Module 1: The basics. [Draft, field testing version.] Association of American Publishers, Washington, DC, 1985.
7. O'Connor M.   1986   International Organization for Standardization: TC46/SC 7 ad hoc group meeting on electronic publishing. European Science Editing 1986; No. 27: p. 10.

Chapter 8

Completing Journal Issues

Before a journal issue can be completed, someone — often a production editor but sometimes the copyeditor — has to prepare copy for the cover and spine; the masthead; the statement of ownership, management, and circulation (in the United States); the contents page or pages; and article identifiers on each page or pair of pages. An index or indexes may be needed for each issue and will certainly be needed for each volume.

Cover

The outside *front cover* of a journal has a standard design, but at least one part, the date, changes with each issue. Use a copy of the previous issue, with dates and other details appropriately amended, to construct complete cover copy each time you prepare an issue for typesetting.

If a complete contents list appears on the back or front cover, double-check that it matches the details in the text as well as matching the list that may also be printed among the first few pages of the issue (see "List of Contents" below). Write "000" in place of any page numbers that are not yet known. If an abbreviated version of the contents list is used on the cover, make sure that this too matches what is in the text as closely as it is intended to.

The outside front cover should carry the eight-digit International Standard Serial Number (ISSN). This code, unique to the journal, should preferably be placed at the top right of the cover, although another prominent place such as the masthead (see below) is acceptable.

Many journals also carry a five-letter or six-letter identifier called a CODEN, preferably also placed on the outside front cover (BMJOAE is the coden for the *British Medical Journal*, for example).

If a journal *spine* is wide enough, it usually carries the journal title or abbreviated title, the date, the volume number, and the page numbers that constitute the issue. Prepare the copy for the spine so that it reads from top to bottom when typed at a right angle to the material on the front cover, though if the journal has a very wide spine the copy can run horizontally from the top, typed in the same direction as the cover copy.

One of the many items that may appear on the inside front or inside back cover is *advertising*. Your involvement with this will probably be limited to seeing that pages containing advertising alone, without any editorial material, are correctly numbered. These pages may be numbered in a separate sequence of small roman numbers (since advertising is often removed before volumes are bound), although some journals use the same sequence of arabic numerals for advertising matter as for the rest of the text.

Instructions to authors are another common choice for the inside front or inside back cover, or this item may be placed elsewhere among the front or back matter. The instructions may appear either in every issue or only in the first issue of each volume. Some journals print a *copyright assignment form* in every issue, for authors to enclose with the manuscripts they submit. Check the copy for the instructions and the copyright form as carefully as you check the rest of the manuscript material for the issue; mistakes seem to creep in even though the same film negative is supposed to be kept for items like this in successive issues.

Masthead

A journal masthead is the place for naming the publisher, the editor, and often the editorial and other staff. It may also state how often the journal is published, how much subscriptions cost, and where correspondence relating to submissions, subscriptions, and advertisements should be sent. The journal's copyright notice and the ISSN (see above) are often included in the masthead. Other information that may be printed here includes the journal's mailing category and a code that identifies the journal if photocopying fees are to be sent to the Copyright Clearance Center or any other fee-collecting society (see "Article Identifiers" below). The masthead is often printed as part of the contents page, where it sometimes shrinks to a line or two in a footnote. In the United States, if the masthead includes the postal notice it should be printed within the first five pages of the issue. Check that the copy is present and correct before the issue is typeset.

Statement of Ownership, Management, and Circulation

In the United States, journals that claim second-class mailing privileges must file an annual statement of ownership, management, and circulation with the Postal Service. This statement must be printed in the journal once a year in a prescribed form and place. On some journals it may be the copyeditor's job to remember when this has to be done and arrange for the copy to be prepared in time. Check the information and mark up the copy as for any other item that is being typeset.

List of Contents

Prepare a list of contents for a journal issue in the journal's standard form, copying the authors' names and the titles of contributions directly from the contributions themselves. Check the list carefully and mark it up as necessary. List everything that the journal usually lists, including—if required—the whereabouts of the instructions for authors, fillers between articles or other items, and lists of corrections if these appear in the issue. Contents lists for the whole volume are sometimes provided in the last issue of a volume or in the first issue of the next volume, but more often indexes are provided instead (see "Indexes" below).

Journal and Article Identifiers, and Running Heads/Footlines

Every journal page or every pair of pages should preferably carry an identifier consisting of the journal title or abbreviated title, the year of publication, and the volume and issue number, e.g., Geol Med 1990; 166(3). Journal identifiers of this kind may be printed as running heads or footlines at the inside margins—on the right for even-numbered pages, on the left for odd-numbered pages. Sometimes the authors' names and short article titles are printed instead of, or as well as, this bibliographical information.

The first page of a journal article (or a chapter in a book) may carry a copyright notice: a C in a circle, the copyright-holder's name, and the date. The copyright-holder may be the author or the publisher, or sometimes a third party. If the publisher wishes to receive photocopying fees from the Copyright Clearance Center (CCC) in the United States, the journal may print its assigned code in the masthead (see above), or the code may be added to the title page of the articles. The copyright notice should be included in this title page identifier, which might take the following form:

Geological Medicine 1990; 166(3):740–764. © ISI Press 1990.
0022-9456/90/166740-25 $01.50/0

The second line in this example consists of the ISSN, the last two digits of the year, the volume number and the first page number of the article, the number of pages in the article, the photocopying fee per copy, and the author royalty or charge per page.

A draft international standard (ISO/DIS 9115)[1] recommends that a code called the biblid (bibliographic identifier or identification), not unlike the CCC code, should be used in journals and in books that have contributions from several authors (see Chapter 10). The biblid provides a unique eye-readable identifier, which could also be machine-readable, for each contribution. Its purpose is to make information retrieval and document ordering easier. Like the CCC code, the biblid is supposed to appear in a prominent place on the first page of each article or chapter. A biblid consists of the ISSN (or the ISBN for books) with the parts separated by hyphens, not spaces; the year of publication in parentheses; the volume and issue numbers; and inclusive pagination for the article or chapter – all printed without any spaces between the characters:

BIBLID 0022-9456(1990)166:3p.740–764

The American equivalent of the biblid is a machine-readable code developed by the Serials Industry Systems Advisory Committee (SISAC); this code has been proposed as an American national standard for serial issue/article identifier code.[2]

If journal practice requires the presence of any of these article identifiers, add them to the title pages of articles before they are typeset; write short titles of articles on the title page too, if they are needed (see above), or supply a list of short titles for running heads if you are sending off all the articles for an issue at the same time. The typesetter should also have or be given a general instruction about running heads or footlines, as discussed at the beginning of this section.

Indexes

As well as the list of contents, an index of authors' names (without the titles of contributions) may be printed in each *issue* of a journal that has more than about 15 to 20 contributions. Some journals print an index of advertisers in each issue too. If indexes of these kinds are your responsibility, check and double-check the spelling and the page numbers, and mark up the index for typesetting in the journal's required style. On the first page of the index copy, write (and circle) the name of the journal and the issue number in which the indexes are to appear.

When a *volume* is complete, most journals provide an author index and a subject index, or a combined author–subject index. There may also be separate indexes for book reviews, editorials, correspondence, corrections, and so on. Making an index is not the subject of this book; if you have to construct one, look for help in the *Chicago Manual of Style*[4] or Butcher[3] or Knight.[5] Checking the copy for the volume index or indexes is much more likely to come your way: check and double-check the spelling, indention, and page numbers, just as for issue indexes, and mark up the copy appropriately for typesetting (see also "Copyediting Indexes" in Chapter 10).

References

1. ISO/DIS 9115. Bibliographic identification (biblid) of contributions in serials and books. ISO, Geneva (in preparation).
2. ANS Z39.56. American national standard for serial issue/serial article identifier code. American National Standards Institute, New York (in preparation).
3. Butcher J.   1981   Copy-editing: the Cambridge handbook, 2nd ed. Cambridge University Press, Cambridge, 1981.
4. University of Chicago Press.   1982   The Chicago manual of style, 13th ed. University of Chicago Press, Chicago, 1982.
5. Knight GN.   1979   Indexing, the art of: a guide to the indexing of books and periodicals. Allen & Unwin, London, 1979.

Chapter 9

Proofreading

Proofreading may be the first job you are given to do as a new copy-editor but it is often done by professionals, whose traditional responsibilities, according to Dellow[1] (p. 4), are:

> to detect and correct all errors of spelling and other typographical errors; to point out any defects of workmanship; to ensure that typographical specifications or other instructions have been followed, or in the absence of such instructions, to ensure, so far as lies within his power, that the printed proof reflects the intentions of the author or customer; to see to it that the 'style of the house' or the customer's wishes, if he has expressed any, have been consistently followed throughout . . . ; to check and correct the punctuation if necessary; to draw attention to any grammatical or possible errors of fact, and to anything which is, or may be, libellous or obscene.

Copyeditors, who take care of some of these activities before manuscripts get into print, may have to follow up their own or someone else's work at the proofreading stage. Even when typesetting is done directly from word processor disks or magnetic tapes, typesetting machines, like human operators — or perhaps because of their operators — still produce mistakes, so at present proofs still need to be checked. But perhaps proofreading will be unnecessary with some systems by the time you read these lines.

Progress of First Proofs

The commonest system of proofing is for a master set carrying correction marks made by the typesetter's proofreader to be sent to the editorial office. A spare set and the original manuscript or printout from the disk or magnetic tape are also sent. Or the manuscript may be returned to the author, with one or two sets of proofs, either direct from the typesetter or via the editorial office. Authors, however, do not always receive proofs; instead they may be asked to check the copyedited manuscript before it is typeset.

If authors do see proofs, they are often asked to return the corrected set to the editorial office, not to the typesetter. The copyeditor then transfers corrections and alterations (or some of them) to the master set and returns this to the typesetter after checking it against the manuscript or printout.

Proofs may be checked in the editorial office before they go to the author, which allows the copyeditor to mark queries for the author's attention. When proofs go direct to the author from the typesetter, the copyeditor can mark queries in brightly colored ink on the manuscript accompanying the proof and call attention to these queries in a covering letter. If the original manuscript is not returned to the author, the typesetter's proofreader may transfer queries from the manuscript to the author's proof before sending it out. If the proofreader doesn't do this, the copyeditor may send the author a list of questions separately and ask him or her to deal with these when the proofs arrive.

Proofs may arrive in two main forms: a first stage of galley proofs on long sheets of paper, and a second stage of paged proofs. Sometimes both stages are in the paged form. Occasionally you may be sent a printout from a line printer — in which case the output looks nothing like the end-product. "Proofs" from a laser printer, however, can look much like the final printed version.

Correcting First Proofs

Whatever form first proofs take, the proofreader works by checking the proof against the original copy word for word, or even character by character; he or she then rereads the proof carefully. Alternatively, the proofreader may read through the whole proof first and check it against the copy afterwards.

Sometimes proof correction is done by two people working together. One person reads aloud from the copy, saying when new paragraphs, indentions, italic or bold type, and so on appear, while the other person looks at the proof and marks the corrections in the ways to be described here. Another variation is for the proofreader to read the text into a tape recorder first and then check the proof while listening to the tape. Scientific and technical material, however, is difficult to check satisfactorily by either of these methods. Proofreading is therefore often a solo job demanding concentration and self-discipline to keep the eyes traveling from manuscript to proof and back, and to stop oneself from falling asleep.

For this occupation you need good eyesight, good lighting, comfortable seating, pens with fine points and different-colored inks, and a smooth, clear working surface at an angle of 15 degrees or so to the top of your desk (the slope is preferable[1] but not essential).

Checking Proof Against Copy

If you are a lone right-handed proofreader, put the manuscript copy in front of you to your left and the proof directly in front of you. Place a ruler or a piece of paper above (or below) the first line of the manuscript (usually the title of the piece), and put the tip of your pen under the first character of the proof. Then, looking from manuscript to proof and back again every dozen words or so, check the proof carefully. The tip of the pen moving forward under each character in the proof helps to stop your eyes from jumping over two or more characters at once and so helps you to spot typographical errors ("typos" in the United States, "literals" in the United Kingdom) and other mistakes, while the ruler or other straight edge makes you focus on the right part of the manuscript. Mark the proof as described below in the section on "Marking the Master Set of Proofs."

Say the words or lines in the manuscript silently to yourself as you check each word and line in the proof: this helps you to catch errors such as plural nouns with singular verbs that may have been overlooked during copyediting or introduced during typesetting.

Check the title, byline, abstract, running heads, headings, and page numbers particularly carefully. Do the same for numerals of any kind in the text, tables, and reference list. Pay special attention, too, to figures, legends, and the remaining parts of tables and reference lists (see section on figures and tables below). Authors checking their own proofs tend to assume that these items are correct, so it is up to the copyeditor-proofreader to pay extra attention to them. Most authors miss typos too, and they seem to turn a blind eye to grammatical errors in their own work — which is why you must watch for these as well, although ideally all mistakes of this kind will have been removed at the copyediting stage.

Make sure that the proof follows the design specifications for type faces, type sizes, and other matters — especially headings and running heads. Check that spacing and indention are as required.

If the proofs are paged, fill in any cross-references to other parts of the document at this stage. Mark any errors in the horizontal alignment of the columns or pages (are columns uneven in length for no apparent reason?).

With typesetting direct from compuscripts, there ought to be very few typos if the earlier stage of copyediting was done well. But other kinds of mistakes may appear instead. If the coding has gone wrong, whole paragraphs may be printed in italics or Greek or gobbledegook; headings may be in the wrong type face or size, or they may have too much or too little spacing above or below them; references may be incorrectly or inconsistently styled; and so on, *ad infinitum*. Ghosts, or dust, may get into the machine . . .

End-of-line word breaks are a particular problem with electronic typesetting. Computers are not as well programmed as human beings when it comes to choosing hyphenation breaks, so words may be divided in less-than-ideal places in proofs. Strictly speaking, breaks are satisfactory if they are divided according to pronunciation (American system, as exemplified in Webster's Third New International Dictionary) or according to derivation (British system). But readers rarely worry unless breaks are "actually misleading or startlingly wrong."[2] (p. 64) — as in "the-rapist," "off-ending," or "geolo-gy" — so it is often better to accept word breaks that are less than ideal rather than run the risk of new errors being produced.

Other proofreading pitfalls to watch for, as Butcher points out[2] (p. 63), include further errors in the same line as an obvious error; misspelled words that form other words ("casual" for "causal"); words that are correct in British (or American) spelling when American (or British) spelling should have been used; and quotations that should have kept their original spelling and punctuation.

Checking Figures, Tables, Legends

Figures (illustrations) nearly always appear in their intended final position in paged proofs, although halftones, the printed versions of photographs, are occasionally provided as separate pulls (proofs) with space left for them in the text. Check that figures in the text appear in the correct order, are attached to the appropriate and correctly numbered legend, and are printed as close as possible to the place where they are first mentioned (see p. 85 and below). See that they are the right way up and the right way round (that is, that they have not become a mirror image of what they ought to be). And make sure that all the original lettering, lines, and numerals are present, legible, and undamaged.

Check the layout: do figures or tables bump into each other? are figures set straight on the page? do figures have squared corners?

Do not ask for figures or tables to be moved elsewhere in the text without a very good reason, such as that they appear in the wrong section of the article. If you measure up what is likely to happen if you ask for such a move, you may discover that the figure or table is already in the most practical place.

Check that legends and table titles correspond to the figures or tables above or below them (depending on journal design). And check, if necessary, that scale magnifications or reductions have been correctly altered if the original figures have been reduced or magnified for printing purposes, or make sure that the original magnification and the photographic reduction are correct ("× 1200, photographic reduction 40%") if both are given.

If figures need to be corrected, return the original figures (the artwork) to the typesetter with the corrected proof. Pack the figures carefully so that they won't get damaged in transit. As well as marking the proof itself, put your corrections or instructions on a separate sheet of paper or on an overlay or photocopy attached to the artwork.

If you are dealing with plates (halftones printed separately from the text), you may have to insert "facing p. 000" below each plate or its legend, or below the first and last of a batch of plates.

Check the numbers in tables carefully against the original copy.

Marking Authors' Corrections

You must guard against any tendency, on your own part or the author's, to allow unnecessary changes on proofs. Corrections take time and cost a lot of money, so you or the editor must decide whether all changes are essential. If authors have approved their copyedited manuscripts, the only changes you should accept from them or from yourself are those essential for scientific accuracy or grammatical correctness. This is not the time to polish the prose or have second thoughts about matters of technical style. If possible, and if necessary, resolve any problems while second proofs are being prepared.

When authors' and other changes *are* acceptable, try to reword sentences — very carefully, without changing the meaning — in such a way that lines with alterations have the same number of characters as the original line. You or the typesetter can ask authors to observe this requirement too; mention it in a covering letter or in instructions on how to correct the proofs. Even with typesetting systems in which page make-up is done automatically, keeping the lines the same length can be more economical than having lines reset unnecessarily.

Finally, transfer the essential changes to the master set of proofs and translate the author's correction marks to standard marks understood by the typesetter.

Marking the Master Set of Proofs (Tables 2 and 3)

You may be asked to use red for marking "printer's" errors (PEs) (actually typesetter's errors) and blue or black for other errors or alterations — whether yours or the author's (AAs); typesetters' proofreaders may use green ink for their corrections and queries on the master set of proofs. This system of color coding allows excess correction changes to be assigned to the publisher or author, one of whom will be liable to pay for alterations cost-

Table 2   American marks for proof correction[a]

Instruction	Mark in text	Mark in margin
Restore deleted characters or words	. - . . . . - . . below characters or words	stet
Remove unwanted marks or reset damaged character	encircle marks or characters	✗
Insert new matter written in margin	∧	the new matter
Delete, or delete and close up	/ or —— through characters or words	℘ or ℘
Substitute character(s) or word(s)	/ or —— through character or words	the new character(s) or word(s)
Wrong font: replace with correct font	encircle character(s) to be changed	wf
Set in italics	—— under character(s)	ital
Set in capital letters	≡≡≡ under characters or words	cap
Set in small capital letters	══ under characters or words	sc
Set initial cap, with small caps for rest	≡ or ══ under characters or words	c & sc
Set in bold type	∿∿∿ under characters or words	bf
Set in bold italic type	∿∿∿ under characters or words	bf ital
Change cap or small cap to lowercase	/ through characters	lc
Change italic to roman type	encircle characters or words	rom
Turn the type right way up	encircle characters	℧
Reset or insert as superscript or subscript	/ through character, or ∧	∨2 /2∧

Table 2   (Continued)

Instruction	Mark in text	Mark in margin
Reset or insert punctuation marks shown	/ through character, or ∧	⊙ : / ; ∧
Reset or insert apostrophe or single or double quotes	/ through character or ∧	∨ ∨∨ ∨ ∨∨
Reset or insert hyphen	/ through character or ∧	=
Reset or insert dash	/ through character or ∧	$\frac{1}{N}$ $\frac{3}{M}$
Begin new paragraph	⁋	⁋
Run paragraphs together	⌐⌐⌐⌐	no ⁋
Transpose characters or words	⌐⌐⌐ around characters or words	tr
Center material	⌐ ⌐ round matter to be centered	ctr
Indent (give size of indent)	∧	□ □ or ☐
Move left	⌐	⌐
Move right	⌐	⌐
Raise	⌐⎯⎯⌐ over matter	⌐⎯⌐
Lower	⌐⎯⎯⌐ under matter	⌐⎯⌐
Correct the vertical alignment	‖	‖ or align
Correct the horizontal alignment	place a line above and below unaligned matter	≡ or straighten
Close up (delete space)	⌒ linking the characters	⌒
Insert space between characters or words	/ between characters or words	#

Table 2 (Continued)

Instruction	Mark in text	Mark in margin
Equalize space between characters or words	✓ between characters or words	eq #
Close up space between lines	) beside lines	) #
Insert space between lines or paragraphs	#	> between lines or paragraphs

a Marks are based on the American National Standards Institute marks.[3]

Table 3   British marks for proof correction*a*

Instruction	Mark in text	Mark in margin
Restore deleted characters or words	- - - - - - - below characters or words	✓
Remove unwanted marks or reset damaged character	encircle marks or characters	X
Insert new matter written in margin*b*	⋏	the new matter ⋏
Delete, or delete and close up	/ or I through characters or ⊢—⊣ or through words	⁊ or ⁊
Substitute character(s) or word(s)	/ through characters or ⊢—⊣ through word(s)	new character(s) or word(s)
Wrong font: replace with correct font	encircle character(s) to be changed	X
Set in italics	——— under character(s)	⊔⊔⌋
Set in capital letters	≡ under characters or words	≡
Set in small capital letters	═ under characters or words	═
Set initial cap, with small caps for rest	≣ or ═ under characters or words	═

Table 3   (Continued)

Instruction	Mark in text	Mark in margin
Set in bold type	‿‿‿ under characters or words	
Set in bold italic type	‿‿‿ under characters or words	
Change cap to lowercase	encircle characters	
Change small caps to lowercase	encircle characters	
Change italic to roman type	encircle characters or words	
Turn the type right way up	encircle characters	
Reset or insert as superscript or subscript	/ through character, or ʎ	
Reset or insert punctuation marks shown	/ through character, or ʎ	(leader dots) (ellipsis points)
Reset or insert apostrophe or single or double quotes	/ through character or ʎ	
Reset or insert hyphen	/ through character or ʎ	
Reset or insert rule	/ through character or ʎ	state size of rule
Begin new paragraph		
Run paragraphs together		
Transpose characters or words	around characters or words (and number them if necessary)	
Transpose lines		
Center material	round matter to be centered	[ ]

Table 3 (Continued)

Instruction	Mark in text	Mark in margin
Indent (give size of indent)[c]	⌐	⌐
Cancel indent	⊢——⌐	⌐
Set line to the measure stated	⊢——⌐ and/or ⌐——⊣	⊢——⊣
Set column to the measure stated	⊢———⊣	⊢———⊣
Move left[c]	⊢—⌐ ⌐ enclosing matter	⌐
Move right[c]	⌐ ⌐——⊣ enclosing matter	⌐
Take over to next line or column or page	⌐——	mark extends into margin
Take back to previous line or column or page	——⌐	as above
Raise[c]	⌐ over matter ⌐—⌐ under matter	⌐—⌐
Lower[c]	⌐—⌐ over matter ⌐ under matter	⌐—⌐
Correct the vertical alignment	‖	‖
Correct the horizontal alignment	place a line above and below unaligned matter	═
Close up (delete space)	◠ linking the characters	◠
Insert space between characters[d]	⎮ between characters	Y
Insert space between words[d]	Y between words	Y

Table 3 (Continued)

Instruction	Mark in text	Mark in margin
Reduce space between characters[d]	| between characters	⌒ |
Reduce space between words[d]	⌒ | between words	⌒ |
Equalize space between characters or words	| between characters or words	⊥
Close up lines	( ) on each side of the column	
Insert space between lines or paragraphs[d]	——(or)——	
Reduce space between lines or paragraphs[d]	——) or (——	

[a] British marks are extracted, with permission, from the British Standards Institution marks in BS 5261: Part 2: 1976,[4] complete copies of which can be obtained from BSI at Linford Wood, Milton Keynes, MK14 6LE (BS 5261C:1976 is a useful eight-page card containing the complete list of 67 marks).

[b] In Continental Europe corrections written in the margins are linked to the appropriate place in the text by one of a series of symbols such as

Γ, F, Ⅎ, L, ⊦, Ⴑ, |, ⌐, ⌀

[c] Give size of indent, rule, etc. when necessary (in ems, ens, millimeters, etc.).

[d] Give size of space or amount of reduction when necessary.

ing more than a certain percentage, perhaps 5% or 10%, of the total typesetting cost.

When you decide that a change of any sort is essential, or if you find a mistake, make two marks very clearly on the proof—in red or in blue or black, as appropriate (see above). Put one mark in the line of text where the change is needed (be careful to mark only those characters that need to be changed, not the ones on either side). Put the other mark, the instruction for the change, in the margin (see Tables 2 and 3 and Figs. 15 and 16). Even though most typesetters can cope well with foreign typemarking systems, try to use the marks that are customary in the country in which or for which you are working.

In the United States and the United Kingdom write marginal corrections or instructions on the left or right of the text, level with the line in which the change is to be made. If a character, word, or phrase has to be

The marks used to ~~delete~~ or insert material in manuscripts (Table) resemble standard proofreading marks, with one major difference in the way they are used. On proofs, all corrections and other changes must be made in the margins as well as in the text (see Chapter 9). In manuscripts, make corrections and changes in or between the typed lines, where the typist or typesetter can see and follow them easily; do not make corrections in the margins unless there is no space left between the lines. Treat a double-spaced or treble-spaced printout from a compuscript in the same way as an ordinary manuscript; but put a cross or other mark in the left margin beside each line containing a change or instruction. If the printout is single-spaced, leaving no room for corrections between the lines, make changes and corrections in the margins and use lines or and loops arrows to show the keyboarder where these changes should be made in the text/ this is easier for many keyboarders than trying to follow a sequence of corrections and marks in the margins.

stet
1

o / e

eq. #

ital
ital

cap
lc

#

o / rom / ∶ = ∶

tr

∶ M ∶

The marks used to delete or insert material in manuscripts (Table 1) resemble standard proofreading marks, with one major difference in the way they are used. On proofs, all corrections and other changes must be made in the margins as well as in the text (see Chapter 9). In manuscripts, make corrections and changes *in or between the typed lines*, where the typist or typesetter can see and follow them easily; do NOT make corrections in the margins unless there is no space left between the lines.

Treat a double-spaced or treble-spaced printout from a compuscript in the same way as an ordinary manuscript, but put a cross or other mark in the left margin beside each line containing a change or instruction. If the printout is single-spaced, leaving no room for corrections between the lines, make changes and corrections in the margins and use lines or loops and arrows to show the keyboarder where these changes should be made in the text — this is easier for many keyboarders than trying to follow a sequence of corrections and marks in the margins.

The marks used to ~~delete~~ or insert material in manuscripts (Table ∧) resemble standard proof-reading marks, with one major difference in the way they are used. On proofs, all corrections and other changes must be made in the margins as well as in the text (see Chapter 9). In manuscripts, make corrections and changes <u>in or between the typed lines</u>, where the typist or typesetter can see and follow them easily; do <u>not</u> make corrections in the margins unless there is no space left between the lines. Treat a double-spaced or treble-spaced printout from a compuscript in the same way as an ordinary manuscript; but put a cross or other mark in the left margin beside each line containing a change or instruction. If the printout is single spaced, leaving no room for corrections between the lines, make changes and corrections in the margins and use lines or and loops arrows to show the keyboarder where these changes should be made in the text; this is easier for many keyboarders than trying to follow a sequence of corrections and marks in the margins.

Figure 16 Same section of text shown in Fig. 15 marked with British proof correction marks.

deleted and new material inserted, use one set of marks, not both deletion and insertion marks (see Fig. 15).

In the United States use an oblique stroke to separate two or more corrections on the same line (but see next paragraph). In the United Kingdom put a long oblique stroke (a longer than usual slash) at the end of each correction mark in the margin. In Continental Europe corrections may be written wherever there is room for them, though they too are usually marked near the relevant line of text, linked to the text by one of a series of symbols placed before the marginal correction, with the same symbol in the text line (Table 3, footnote *b*).

If several changes have to be made in a line or successive lines, delete the whole section and write the correct version in the margin in full; this is easier for the typesetter than if you make a string of small changes. Do not write corrections up the sides or on the back of the proof; this kind

Figure 15 Section of text marked with American proof correction marks and corrected proof.

of marking slows up the typesetter's work and is therefore more expensive than if you wrote the corrections properly. If long passages have to be inserted or substituted for incorrect passages, type them on a separate slip of paper and tape one side of the slip to the page of proof to which it refers — don't staple or pin it or use a paperclip. Link the insert to its position in the text by using an insert mark (caret) in the text, with a message or mark in the proof margin: "Insert 1 attached," for example, and a corresponding message ("Insert 1 for p. 99") on the typed slip; in the British system[3] put a diamond with a letter inside it (⟨A⟩) on the proof, with the same symbol on the insert slip. Make sure that the page number or any other necessary identification appears on the insert slips in case they go adrift.

Ideally the corrections on the master set of proofs should be copied onto a spare set before the master set goes back to the typesetter. Follow office practice on this, whatever it may be.

Checking Second Proofs

Authors of journal articles rarely receive a second set of proofs to check. Someone outside the typesetting house, however, usually checks that all the corrections and changes marked on the first proofs, whether they are galleys or paged proofs, have been correctly made and that no new mistakes have crept in. If this job falls to your lot, check the whole line and the lines before and after it when even a small change has been made; even with electronic typesetting, extra mistakes sometimes creep in when corrections are being made.

Check again that figures are the right way up and the right way round, and have the right legends printed with them.

Make sure that running headlines or footlines, if any, are correct, and that page numbers run consecutively. If you are dealing with proofs of a complete issue of a journal, check that the page numbers run on correctly from the previous issue and that page numbers in the contents list match the numbers on the first pages of the articles and other items, or insert the page numbers in the contents list now, as necessary.

If you are responsible for a complete issue of a journal, you may need to proofread contents lists and indexes, as well as the main articles and other features.

Lastly . . .

Don't allow your proofreading to sink to the depths remarked on by this book reviewer:[5] "Misprints abound, dates are reversed, brackets unclosed, whole lines are missing, and to cap it all the footnotes do not synchronise with the text." But don't be overcome with mortification if you later find a mistake or two in the printed journal or book: absolute perfection is rare.

References

1. Dellow EL.   1979   A first course in proof correcting. Northgate, London, 1979.
2. Butcher J.   1981   Copy-editing: the Cambridge handbook, 2nd ed. Cambridge University Press, Cambridge, 1981.
3. ANSI Z39.22-1974 American National Standard proof corrections. American National Standards Institute, New York, 1974.
4. BS 5261-1976. Guide to copy preparation and proof corrections. Part 2: Specifications for typographic requirements, marks for copy preparation and proof correction, proofing procedure. British Standards Institution, London, 1976.
5. Morrogh M.   1985   [Book review]. Irish Times 1985 (17 August).

Chapter 10

Copyediting Books — The Extra Tasks

Copyeditors of books work in much the same way as copyeditors of journals. Both kinds deal with language, mechanical style, and format, with or without substantive editing, as described in Chapters 2–7 (just substitute "book" or "house style" for "journal" when you read those chapters). Full-time copyeditors in publishing houses may also brief estimators and designers[1] (p. 5–15), set up production schedules, and help to launch promotion plans. If the last three activities fall to your lot, plenty of in-house advice should be available (and see Refs. 1, 2, and 3). In particular, inquire what the designer's specifications mean and follow them carefully.

This chapter describes different kinds of books and the problems specific to them, then discusses chasing and guiding contributors and preparing front and back matter.

Dealing with Different Kinds of Books

The books you copyedit in the ways already described may be textbooks, monographs, conference proceedings, or symposia. Monographs, like textbooks, may have one author, or several or many, while conference proceedings and symposia obviously always have many authors. These categories overlap: symposia may be classified as monographs; some monographs may become textbooks. And note that "single-author" as used in this chapter really means either one or several authors (up to about six) in a book that has no named editor.

Textbooks

Every kind of book, of course, demands and (one hopes) deserves great care and attention on the part of the copyeditor, but textbooks call for the greatest possible accuracy at every level, from statements of fact down to spelling and punctuation. If you work on textbooks, triple-check every detail, especially on the artwork (usually created in the publishing house or by freelances rather than provided by the authors). You will probably work closely with the authors and will need to discuss many points of detail with them. On top of your regular copyediting tasks you may need to remind the authors who their readers will be, if they seem to be writing for the wrong audience.

Some of the "don't's" laid down for journal copyeditors do not apply to textbook copyeditors. For example, some repetition is essential in lengthy textbooks; terms with abbreviations may be used in full from time to time, even after the abbreviations have been used and explained earlier; figure legends are helpful to readers if they include descriptions such as "Electron micrograph of . . ."

Monographs

A monograph is a "separate treatise on [a] single object or class of objects" (*Concise Oxford Dictionary*), not necessarily a book written by one author. For the multi-author kind, good guidelines for authors are a help, as discussed later in this chapter. The section on "Chasing Authors" should be useful too.

Conference Proceedings

Chasing authors will certainly be relevant if you are copyediting the proceedings of a conference, especially if the organizers did not decide on publication until after the meeting was over, with the result that the speakers produce poor manuscripts or none at all. Conference proceedings published 18 months or more after the meeting was held are already well out of date, so your first priority should be to get the papers into print (or perhaps into a full-text database) as quickly as possible, even if you have to skimp on some questions of detail.

Ideally, editors of conference proceedings should be members of the organizing committees from the beginning[4] (p. 109) and be able to draw up a set of instructions for authors, including a firm timetable for the receipt of manuscripts. If manuscripts are received before or during a small or medium-size meeting, the editor and copyeditor can even arrange to edit them while the meeting is still going on.[5,6] It is difficult to do on-site editing at large meetings, but the principles are the same: get your hands on as many manuscripts as possible as soon as possible, edit/copyedit them as quickly as possible, and send them on their way to the typesetter as promptly as possible. If you attend the meeting yourself (which is to be recommended), try to establish firm dates for receipt of the missing manuscripts.

The instructions for authors should ask for manuscripts written for publication, not notes or other versions of the oral presentation. If the latter kind is what you get, try to convert contributions into manuscripts more acceptable for publication: remove the chatty asides; condense the wording; ask for abstracts, tables, figures, and references to be supplied if these have not been provided. Make sure that the title is appropriate (matching the contents, informative), that the authors and their addresses are correctly given if the presenter has co-authors, and that the author or authors have signed a copyright form or license to publish, if so required.

Symposium Proceedings

Symposia are (or ought to be) mini or micro versions of conferences, with fewer speakers and more chance of getting the book out quickly. Again, the manuscripts should be written for publication, not just for presentation at the meeting, and good guidelines for authors are essential. If manuscripts turn out to be presentation versions, deal with them as described above.

The extra ingredient in symposia is often the discussion. If these sessions are tape-recorded for inclusion in the publication (though there are methods other than tape-recording for dealing with discussions), arrange to have treble-spaced transcripts with wide margins supplied. Take notes of speakers' names during these sessions, or get someone else to do so, so that you can be sure that comments are attributed to the right people. Listen to the recordings yourself and edit the transcripts into a reasonable and readable version of the original conversation, insofar as the clarity or otherwise of the recording allows.

Editing discussion sessions includes rearranging comments in a more logical order than in real life (readability is nearly always more important than a list of comments in their original order); removing or condensing rambling comments; removing polite (or impolite) comments of no interest to readers who were not at the meeting; and noting where extra information, including bibliographical references, should be supplied. When you have done all this, have the edited transcript retyped, still in treble spacing. Write your questions and requests for information in the margins and send speakers their comments in context; at a minimum send the page before and the page after each page containing comments by the person you are writing to. Tell speakers how to correct the edited transcripts — for example, how to present references. Set a reasonable but early deadline for the return of corrected transcripts and start chasing up the laggards as soon as the deadline date is reached.

Next, put the jigsaw back together again. Make sure that the comments still flow logically, that your queries have been answered, that contradictory statements have been reconciled, and that speakers' questions to each other have been answered or deleted, or — if appropriate — reworded as statements instead of questions. Prepare reference lists if these are needed and, finally, read and edit the articles and their discussions together. Always keep the readers in mind: that is, expand jargon into generally understood language, spell out abbreviations if there are too many of them, or at least explain them at first mention, and insert cross-references where they are needed[4] (p. 115–118).

Chasing Authors

One of your first jobs on a book of any category may be to chase up dilatory authors for their contributions, or help the editor to chase them. Authors of monographs may need encouragement to meet their deadlines; contributors to multi-author books may need something stronger than encouragement. You (or the editor) may have to stage a gradually intensifying barrage of letters and telephone calls, backed up by telex, facsimile, or computer network messages — cajoling, demanding, and threatening. The barrage may work up to threats to drop contributions, or publish unedited transcripts of tape-recorded presentations (if available), or ban any further contributions from those authors in their lifetimes. This kind of blackmail is not always effective; scientists are often more interested in getting their articles accepted for publication in primary journals than in preparing chapters for books. But if an author's contribution is essential to the success of a textbook or monograph, you must be persistent. Production of a book rarely begins until all the contributions are on the publisher's desk, and one recalcitrant author can delay publication for months, sometimes years.

Guiding Authors

One way of encouraging authors to submit satisfactory manuscripts on time is to provide them with clear guidelines or instructions at an early stage. Publishing houses usually have their own guidelines for single-author as well as multi-author books — but if you are a freelancer or an authors' editor you may sometimes work with scientists who have never edited a book before and whose publishers have not provided any instructions for authors. In this event, even if there are only a few contributors, draw up your own version based on any suitable guidelines you can lay hands on. The editor will have outlined some points in letters to potential authors but it is useful if these are repeated in the guidelines, for easy reference.

For a textbook or monograph begin the guidelines with a short description, written by the editor, of the aims and scope of the book and the kind of readership it is intended for. Then tell authors who the publisher is, what (if anything) their fee or share of the royalties will be, and whether they will be given reprints of their chapters, or a copy of the book, or both. Explain that they will be asked to sign a contract or a copyright assignment form when they accept the editor's invitation to contribute to the book. Remind them that they will need written permission from copyright-holders if they want to reproduce borrowed material in their chapters. For the benefit of these borrowers, say whether world English-language rights or more restricted rights to the borrowed material should be requested, and mention how many copies of the book will be printed (if known) so that these pieces of information can be included in their permission requests (some publishers base their decision on whether to charge a copyright fee on the print run, among other matters).

Include a realistic timetable for the receipt of manuscripts or printouts from disks/magnetic tapes (compuscripts) and for the arrival of proofs. Tell contributors how long their chapters or articles should be and how many copies of the manuscript or printout will be needed (editors and copyeditors usually do their preliminary work, at least, on hard copy). State which reference style is to be used (see "Dealing with Cross-references and References" below); say how many tables and figures may be included; and explain how tables, figures, and any other parts of the chapter or article are to be prepared (describe the style and format).

Remind authors to number the pages of manuscripts or printouts and to separate fanfold paper. Ask them to use a good-quality printer, not one with type lacking descenders, and to leave the typed lines unjustified on the right (ragged right).

If the publisher's typesetter accepts disks or magnetic tapes, tell authors which formats are suitable and whether and when any codes should be keyed in, or ask them to tell you, before they start keyboarding their contributions, which brands of hardware/software they will be using. Tell them whether they or the typesetter will be expected to enter the final changes and corrections.

Keeping Track of Progress

Keep track of contributions to multi-author or single-author books by drawing up a chart listing the articles or chapters in their expected order of appearance in the book (the running order). Assign reference numbers to the manuscripts/printouts even if chapter numbers will not be printed in the book. Include columns for the date the manuscript or printout is received, and perhaps also for the number of pages, number of tables and figures, and so on. When a manuscript reaches you, fill in the relevant columns on the chart and write the reference number on the title page, on the tables, figures, and legends, and on any other parts that might become separated from the main text during typesetting. For a compuscript, write the reference number on the title page, only, of the printout.

Calculating Length

You may be required to estimate the length of contributions, or the whole book, by doing a *cast-off.* Count the number of characters on a typical page of the manuscript, divide the total by the known character count per pica for the type face and size in which the book is to be set, then divide by the number of picas per line, as decided by the designer. This will give you the number of printed lines each manuscript page will make:

1500 characters ÷ 3.0 characters per pica = 500 picas
500 picas ÷ 25 picas per line = 20 lines

A further sum based on the number of manuscript pages and the number of lines on the printed page (also decided by the designer) will show the approximate number of pages in the book:

(20 lines × 300 manuscript pages) ÷ 40 lines per printed page = 150 printed pages

Count footnotes, legends, and other such material separately (see Chapter 5), and estimate how much space figures and tables will occupy.

Adding Identification Codes

The title pages of chapters in books, like those of articles in journals, may carry the Copyright Clearance Center (CCC) code or a bibliographical identification code (biblid) (see Chapter 8). The biblid for a contribution to a multi-author book consists of the International Standard Book Number (ISBN) or the International Standard Serial Number for books in series, if the latter have no ISBN; the year; and the inclusive page numbers of the article or chapter, printed without spaces betwen the different elements:

BIBLID 0-471-04932-8(1979)p.40–65.

If either the CCC code or the biblid is to be used, put the code in the place specified by the designer at the beginning of each article, or indicate that the information will be supplied in proof.

Numbering Tables and Figures

A minor difference between books and journals is that tables and figures may be numbered consecutively through a book rather than article by article, but this depends on the type of book. Check whether the design calls for arabic or roman numerals for the tables and figures.

Dealing with Cross-references and References

Another matter to be decided is how to deal with cross-references: if there are a lot of these, can chapter or section numbers be used instead of leaving zeros in the copy? Inserting page numbers in proof is expensive.

Decisions have to be made about bibliographical references too. The wonderful variations in reference style that are promulgated in journals (Chapter 6) also flourish in books. For multi-author books that are not in a series the editor may decide to follow the publisher's house style. Details of this style should be included in the guidelines for contributors (see above). If the editor decides that consistency throughout the book can be dispensed with, tell authors (in the guidelines) what is acceptable and ask them to be consistent within their chapters. Multi-author books in an established series will already (one hopes) have a consistent reference style or set of styles. If a new publisher takes over an established series, the style or styles should stay the same unless the editor decides it is time for a change; imposing an unfamiliar house style can waste a lot of time for everyone, not least the copyeditor. For the same reason, publishers usually accept the reference style chosen by the author of a single-author book, provided it is consistent and provides enough information for readers who want to find the original sources.

Instead of a reference list at the end of each chapter, some multi-author books have a consolidated reference list at the end of the book. The job of consolidation will almost certainly fall to you as the copyeditor, as will the work of renumbering references throughout the book, if numbers are being used. Allow plenty of time for this.

Preparing Front Matter (Preliminary Pages)

When the articles or chapters for a book have been copyedited in the ways described in Chapters 3–7, prepare the front matter or preliminary pages (prelims) for typesetting in the form decided by the designer. You may be required to number these pages — the half-title page, title page, copyright page, and so on (see below) — in lowercase roman numerals (i, ii, iii, iv, etc.), even if arabic numerals are to be used in the printed book. (See Refs. 1 and 2.)

Half-title Leaf

The *half-title* leaf carries the title of the book on the front (recto; p. i) and usually nothing else; if the back (verso) is to be blank, write "p. ii blank" on p. i. The verso may be used to list the series editors, or other books in the same series, or the authors' names if there are too many to fit on the title page. An author's previous publications or information about the organizers or sponsors of a symposium are other items that might be listed on p. ii. If a frontispiece is placed here or elsewhere among the prelims, make sure that permission has been obtained to reproduce it, and prepare a legend crediting the photographer or artist as well as the copyright-holder ("Charles Brown, 1900–1990. Photograph by John Smith, reproduced by permission of the Jones Institute, New York, NY").

Title Page

As well as the title of the book, with subtitle if there is one, put the author's or editor's name on the *title page* (p. iii), with the series name if there is one, and the place and date of the meeting for conference proceedings (or the place and date may go in another prominent position in the prelims, such as at the head of the list of contents). For layout, follow the designer's plan. At the bottom of the page type the publisher's name, the city or cities of publication, the edition number if it is the second or later, and, if required, the date (year) of publication.

Copyright Page

The verso of the title page is the copyright page (p. iv), where you should give:

1. the publisher's name and full address(es)
2. the copyright notice (©, with the date or dates of successive editions and the name of the copyright-holder; add "All rights reserved" for books for which protection under the Buenos Aires Convention is required[2] (p. 7)—and similar or longer messages may be added, depending on the country of publication
3. the publishing history ("Published 1986. Reprinted with corrections 1987, 1988. Second edition 1989")
4. the country [and perhaps the place] where the book is to be printed ("Printed in the United States of America [at the Bath Press, Bath, CN]")

5. the International Standard Book Number (or the International Standard Serial Number, for some books), if known — or write "ISBN" or "ISSN" and indicate that the number will be supplied later.

"Cataloging in Publication" (CIP) information, obtained by the publisher from the Library of Congress, or from the British Library or other national center, is usually added to the copyright page while typesetting is in progress, or when the book is in proof; write "Cataloging in Publication data" on the copy or proof, if necessary. CIP data have to be set exactly as the Library supplies the information; if there are any mistakes on the data card received by the publisher, a revised version has to be requested from the Library.

Item 4 above is sometimes expanded into a *colophon*, or description of the production of the book. This may include the designer's name, the typesetter's name, the type face used, the type of paper, and the type of binding, as well as the printer's name or name and address. The colophon may instead be printed on the last page of the book. The word colophon (literally the summit, or finishing touch) is occasionally also used for the publisher's device which appears on the title page and spine.

Prelims After the Copyright Page

The copyright page may be followed by a page or pages carrying a *dedication* or an *epigraph* (a quotation relevant to the subject of the book, with the name of the quotation's author and the title of the work it comes from: not often used in scientific books). The verso of a dedication, if not used for an epigraph, is usually left blank.

After p. vi, or after p. iv if there is neither dedication nor epigraph, the standard order is (as needed): contents, lists of illustrations and tables, list of contributors, foreword, preface and acknowledgments, introduction (if this is not part of the main text), and list of abbreviations. The first page of each of these, except the list of abbreviations, is usually a right-hand page, unless the design calls for a different arrangement; mark the copy accordingly.

The *contents page or pages* should list all the front matter after the epigraph (if there is one); the main contents; and the back or end matter (see below). Type the chapter titles, and the authors' names for multi-author books, from the final versions of the manuscripts or printouts and check the details carefully. If the design calls for subheadings within chapters to be included in the contents list, type them according to the design — usually they are indented beneath the chapter title. Authors' positions ("Chairman, Department of Geology, University of South Columbia") may also be included. Put "000" for page numbers — unless these are already known, of course.

Prepare a *list of illustrations* and a *list of tables* in the same style as the contents list. Short versions of the figure legends or table titles are sufficient, with page numbers again indicated by "000." These two lists, however, are rarely included in multi-author books, and are not needed if there are very few illustrations, or if there are a lot of charts and graphs that are closely linked to the text[2] (p. 17).

A *list of contributors* will almost certainly be needed for multi-author books, although if there are fewer than 15 authors they may be listed on p. ii, opposite the title page (see above). Arrange this list alphabetically, except when there is a particular reason for some other order to be used. Authors' addresses or short addresses may form part of this list, and brief biographies may be printed here, if needed, or these might be included in the back matter.

A *foreword* is generally written by someone other than a contributor to the book; a *preface* may be by the author or editor, explaining how the book or the meeting on which the book is based came about. *Acknowledgments* are often included as part of the preface, but if they consist of a list of sources of borrowed items this is usually printed separately. Another place for acknowledgments is at the back of the book, before the index. Copyedit these items and mark them up according to the requirements of the design.

A *list of abbreviations*, if included, is usually printed on the last left-hand page before the text[1] (p. 131–132). When you prepare this list, check that it is in proper alphabetical order, that it is complete, and that it agrees with the usage in the text.

Preparing Back or End Matter

The back matter may include appendixes, notes, a glossary, a list of references cited or a more extensive bibliography, a list of further reading, and an index or indexes, as well as brief biographies of contributors, the acknowledgments, and a colophon, if these are not placed with the front matter (see above). For single-author books, or books by only a few authors, the authors will supply most of these items at the same time as the rest of the material, or soon after (with luck). Authors sometimes also prepare their own indexes when the book reaches the proof stage, or indexes may be prepared by a professional indexer. Automatic computer indexing during typesetting is sometimes used, but human beings are still better than machines at producing satisfactory indexes. In multi-author books any appendixes, notes, and bibliographies usually appear after the relevant chapters rather than at the end of the book, leaving just the index or indexes, whoever or whatever prepares it, to be dealt with when page proofs are ready.

Correct the spelling, punctuation, and grammar of all these items, where necessary (and see "Copyediting Indexes" below). Check for consistency in capitalization, italicization, and other aspects of mechanical style, and mark the format for the typesetter (see Chapters 5–7).

Adding the Final Touches

When you have copyedited the end matter other than indexes, as listed above, you will be almost ready to send the material off for keyboarding if it is in manuscript form, or for final corrections and coding if it is held on disks or magnetic tapes. Before you parcel everything up, however, there is more to do.

Listing Running Heads and Footlines

The typesetter usually needs a list of running heads or footlines for the whole book, from the contents page(s) onwards, including the yet-to-come index or indexes. For multi-author books the question is whether to have authors' names on the left (verso) with a short chapter title on the right (recto), or whether the book title should appear on the left all the way through, with authors' names or short chapter titles on the right. And so on. Or should there be any running heads/footlines at all?

If head or footlines *are* to be used, what is their maximum length? Are they to be set in small capitals, in uppercase and lowercase italics, or in any other style?

For single-author books the questions may be: book title left, chapter heading right, or chapter heading left and a right-hand heading that changes with each section of the chapter? Or, again, are running heads/footlines needed are all?

Prepare the list in two columns, for verso and recto pages, following whatever plan has been laid down by the designer. If running heads or footlines are to change with each section within chapters, draw the typesetter's attention to this and type a list of all the possible running heads, since if some sections are short they won't all need a heading, and you won't be able to predict exactly which running heads will be used.

Numbering the Pages

The publisher or typesetter may ask for the manuscript copy or the printouts to be numbered all the way through. Use roman numerals for the preliminary pages (see above) and arabic numerals for the text, starting with p. 1 for the text however many prelims there are. Include in your numbering the title pages of chapters or articles, reference lists, legends for figures, and any other text matter there may be, but not tables or figures.

If the book is to be produced from author-prepared camera-ready copy, or if the copy is to be scanned by an optical character recognition machine (OCR), write the page numbers outside the main typing area and keep the pages as clean as possible.

Copying and Transmitting the Manuscript/Compuscript

Most publishers ask for two sets of the complete manuscript — one for the typesetter and one for the designer or estimator — and you should keep a copy in the editorial office if you work anywhere other than in a publishing house. Make the copy or copies after you have numbered the pages (see above). If there is a "ribbon" or top copy, send this to the publisher (or to the production department or typesetter, if you work for a publisher), with the other copy or copies needed. Enclose a transmittal sheet: a list of what is being sent, what has still to be sent (and some indication of when it will be sent), the number of sets of proofs needed (and when), and any special instructions you may have for the typesetter (see Ref. 2, p. 78–80).

Preparing Cover or Jacket Copy

The cover or jacket copy should be prepared and sent to the publisher or typesetter with or shortly after the manuscript or compuscript. The designer specifies what is to appear on the front, back, and spine, or for a series there may be a well-established design which you can copy from an earlier book and modify as necessary for the volume you are working on.

The front cover carries the title, subtitle (usually), and the authors' or editors' names. The spine carries the title, the authors' or editors' names, the publisher's name, and, occasionally, the publisher's logotype or device (sometimes called a colophon: see above). The back cover or the jacket flaps may include a blurb, information about the authors or editors or sponsors of meetings, the titles of other books in a series, and similar information. The back cover or jacket should also carry the ISBN or ISSN number (or both), and in some countries a bar code is printed here too (see below).

Writing Blurbs

In trade (general) publishing, the publicity or marketing specialists write the blurbs for book jackets, catalogs, mailing leaflets, and so on. In scientific, technical, and medical publishing, or indeed for any scholarly book, copyeditors or editors or the authors may find themselves acting as blurb writers (copywriters). If you have to write the blurb for a multi-author book — and who is better acquainted with the contents than you? — ask how many words are needed. In a 200-word blurb you might use the first two sentences to explain the need for the book or the reason why a conference or symposium was held. You could state the theme or purpose of the book in the next sentence or two, summarize the contents in four or five sentences, and end by naming the likely readership if this isn't already clear. If a much shorter blurb is needed, leave out the summarizing sentences and modify the rest appropriately (Ref. 4, p. 74–75; Ref. 7, p. 211–218). Whoever writes the blurb, check that it matches the text in substance and spelling.

ISBN/ISSN and Bar Codes

The International Standard Book Number or the International Standard Serial Number, or both (for books in series), should appear at the bottom of the back cover or jacket as well as on the copyright page. Books printed in the United Kingdom and elsewhere in Europe should carry a machine-readable version of the ISBN printed close to a bar code representation of the European Article Number (EAN). Books published in the United States may carry a bar code representation of the Universal Product Code (UPC) on the back cover or jacket, or both. (The UPC code in the United States is now scheduled to give way to "Bookland EAN".) In preparing the cover copy, indicate where the bar code should be placed and, if required, provide the printer with the necessary piece of film for the code.

Copyediting Indexes

A subject index is the commonest kind of index included in scholarly books, with an index of contributors often added in multi-author books. An index of names of people referred to, or of places, drugs, and so on, may sometimes be required. Whatever the kind of index, check the copy (sometimes provided on paper slips or cards) as carefully as you check everything else for the book.

First, some substantive editing may be needed. If there are large numbers of subentries under a main heading that is similar to the subject of the book (e.g., "copyediting"), or if there are either long strings of page numbers after main entries or only one page number beside each subentry, the indexer may not have done the job well. If you copyedited the book whose index you are working on you may also recognize some entries as being too trivial to be included, or you may realize that some important items have been left out. If you know the book well, you can remedy the last two defects yourself, but think carefully before tackling those first problems: once you start changing entries you may upset the whole structure of the index. If you think there is something badly wrong, consult your editor.

Then check whether all the entries are spelled correctly, or at least spelled in the same way as in the text (make spot checks in the proofs). Are the main entries correctly alphabetized (numbers, Greek letters, and some Latin prefixes should be ignored in putting an index in order)? Are subentries and sub-subentries correctly alphabetized? Are they appropriately run-in or indented, depending on the style specified by the designer? (In the "run-in" style subentries follow on from each other, separated by semicolons or dashes; in the indented style each subentry starts on a new line.) Is the punctuation correct? Do the page numbers make sense ("99–90" and "56, 34, 98" are nonsense) and are they correct (make more spot checks in the text)?

When you have copyedited the index in this way, mark it up in the required style for the typesetter. Write the title of the book on the first page of the index copy and say which page number the index, or the first index, should begin on. Say which type face and type size are to be used, if this is one of your responsibilities, and mark indentions if the indented style is being used (e.g., one en or one em for subentries, two for sub-subentries, and three for runover lines). If extra space is needed between entries beginning with different letters of the alphabet, or if entries are to be set in bold or in capitals or any other distinctive way, mark the copy accordingly.

References

1. Butcher, J.   1981   Copy-editing: the Cambridge handbook, 2nd ed. Cambridge University Press, Cambridge, 1981.
2. University of Chicago Press.   1982   The Chicago manual of style, 13th ed. University of Chicago Press, Chicago, 1982.
3. Judd, K.   1982   Copyediting: a practical guide. Kaufmann, Los Altos, CA, 1982.
4. O'Connor, M.   1978   Editing scientific books and journals: an ELSE–Ciba Foundation guide for editors. Pitman, London, 1978 [published in the United States as The scientist as editor. Wiley, New York, 1979].
5. (Dickens B).   1979   The technical editor's view. As reported in Earth & Life Science Editing 1979; No. 8: p. 6–7.
6. (Franklin P).   1984   Publication of proceedings. As reported in Earth & Life Science Editing 1985; No. 23: p. 7.
7. Bodian N.   1984   Copywriter's handbook: a practical guide for advertising and promotion of specialized and scholarly books and journals. ISI Press, Philadelphia, 1984.

Epilogue: Copyediting Concluded

An encyclopedia, or perhaps an encyclopedia for every branch of science, would be needed to answer all the questions copyeditors must cope with. In this short book I have tried to explain the general principles of copyediting in science and point to ways of dealing with the main problems. The various books and articles referred to will give you much of the extra information you'll need on grammar, nomenclature, mechanical style, format, and so on. I'll leave you with a list of Edicts of Copyediting to contemplate, including some not yet mentioned (Edict No. 1 appears in Chapter 1, Nos. 2–7 in Chapter 2). You'll probably have a few variations of your own to add after a few months as a copyeditor.

1. Leave well enough alone.
2. Find good authority for every change.
3. Make essential changes only. Ask yourself
   a. Is a change really necessary here?
   b. Why is it necessary?
   c. Is my version an improvement — or a backward step?
4. Protect readers from authors and authors from themselves.
5. Argue so far and no further; then give way gracefully.
6. Respect authors' feelings — praise before you criticize.
7. Always comment constructively.

To which should be added:

8. Keep messages on manuscripts short and to the point.
9. Try to find the answer yourself: look it up.
10. Be consistent: follow house style (or invent your own).
11. If other guidance on format and style is lacking, look at recent issues of the journal or at recent books in a series.

And a last thought:

12. Enjoy yourself: copyediting is hard work but it has its rewards . . .

Appendix: Useful Addresses

The addresses given here were correct in mid-1986, to the best of my knowledge. Presidents and Secretaries of associations (part 1), however, have a high turnover rate: you may need to do some detective work as time goes by. (Information about changes and possible additions, especially on courses in editing/copyediting, would be welcomed.)

(1) Associations and other organizations of interest to editors and copyeditors in science and medicine [with the titles of their journals/bulletins/newsletters] *

African Association of Science Editors (AASE). President: Professor T. Odhiambo, International Centre of Insect Physiology and Ecology, P.O. Box 30772, Nairobi, Kenya; Secretary: K. S. A. Buigutt, Kenya Rangeland Ecological Monitoring Unit (KREMU), P.O. Box 47146, Nairobi, Kenya [*AASE Bulletin*]

American Medical Writers Association (AMWA). Executive Administrator: Ms Lillian Sablack, AMWA National Office, 5272 River Rd., Suite 370, Bethesda, MD 20816, USA [*Medical Communications*]

Association des Rédacteurs et Editeurs Scientifiques d'Afrique Francophone (ARESAF). In process of formation (contact Dr Abdoul Azziz Ly, CODESRIA, P.O. Box 3304, Dakar, Senegal)

Association of Earth Science Editors (AESE). Secretary/Treasurer: Hal L. James, Montana Bureau of Mines and Geology, Butte, Montana 59701, USA [*Blueline*]

Association of Learned and Professional Society Publishers (ALPSP).

---

* Meetings of several of the associations listed here, such as the American Medical Writers Association, the Association of Earth Science Editors, the Council of Biology Editors, the European Association of Science Editors, and the International Federation of Scientific Editors' Associations, include lectures and workshops on subjects of interest to copyeditors.

Secretary: A. I. P. Henton, Sentosa, Hill Road, Fairlight, East Sussex, TN35 4AE, UK [*ALPSP Bulletin*]

Center for Book Research, University of Scranton, Scranton, PA 18510, USA. Director: John P. Dessauer [*Book Research Quarterly*]

Collège Français d'Enseignement de la Communication Scientifique (COFECOS). President: Professor Roger Bénichoux, Institut de Recherches Chirurgicales, CHU de Brabois, F-54511 Vandoeuvre-les-Nancy Cédex, France [*Editologie*]

Commission of Editors of Journals Concerned with Clinical Chemistry (CEJCCC)†: c/o P. M. G. Broughton, Wolfson Research Laboratories, Department of Clinical Chemistry, University of Birmingham, B15 2TH, UK

Committee of Editors of Biochemical Journals (CEBJ).† Chairman: Dr Herbert Tabor, Journal of Biological Chemistry, 9650 Rockville Pike, Bethesda, MD 20814, USA; Secretary: Dr Chris Pogson, Biochemical Journal, 7 Warwick Court, London, WC1R 5DP, UK [circulates new recommendations on nomenclature to "Interested Editors" as well as to the Full and Associate Members of CEBJ]

Council of Biology Editors (CBE). Executive Director: Philip Altman, Council of Biology Editors Inc., 9650 Rockville Pike, Bethesda, MD 20814, USA [*CBE Views*]

Editorial Experts Inc., 85 South Bragg, Alexandria, VA 22312, USA, tel. 703 642-3040 [*Editorial Eye*]

Editors' Fellowship of Academic Societies and Associations: c/o Tadahiro Ohmi, Center for Academic Publications, 4-16, 2 chome, Yayoi, Bunkyo-ku, Tokyo 115, Japan

Editors of European Chemistry (EdEuChem)†: c/o Professor Erno Pungor, Institute for General and Analytical Chemistry, Technical University, Gellert Ter 4, 1502 Budapest XI, Hungary

European Association of Science Editors (EASE). Secretary-Treasurer: Miss Nadia Slow, 18 St Paul's Square, Bromley, Kent, BR2 0XH, UK [*European Science Editing*]

Finnish Association of Science Editors and Journalists: see Suomen tiedetoi-mittajat, below

IEEE Professional Communication Society, Institute of Electrical and Electronics Engineers, 345 East 47 Street, New York, NY 10017, USA [*IEEE Transactions on Professional Communication*]

International Association of Anthropology Editors (IAAE): c/o Professor Cyril Belshaw, Current Anthropology, University of British Columbia, 6303 N.W. Marine Drive, Vancouver, B.C., Canada V6T 2B2

---

† CEJCCC, CEBJ, and EdEuChem are associations specifically for scientist-editors in clinical chemistry, biochemistry, and chemistry; they do not include copyeditors among their members.

International Federation of Scientific Editors' Associations (IFSEA). Secretary-General: Dr Elizabeth M. Zipf, BioSciences Information Service, 2100 Arch Street, Philadelphia, PA 19103-1399, USA

Netherlands Association of Science Editors (NASE): see Wetenschappelijke-Redacteurenkring, below

Nordic Publishing Board in Science (NOP-N). Secretary: Ms Karin Westerlund, NOP-N, Finlands Akademi, Banmästargatan 12, 00520 Helsingfors 52, Finland [*NOP-Nytt*]

Scholarly Publishers Association, Singapore

Singapore Society of Editors

Society for Biomedical Communicators (India) (SOBIC). c/o Dr G. V. Satyavati, Senior Deputy Director-General, Indian Council of Medical Research, P.O. Box 4509, Ansari Nagar, New Delhi 110029, India

Society for Scholarly Publishing (SSP). Administrator: Ms Alice O'Leary, Society for Scholarly Publishing, 2000 Florida Avenue NW, Washington DC 20009, USA [*Letter*; members also receive *Scholarly Publishing*]

Society for Technical Communication, 815 Fifteenth Street NW, Washington, DC 20005. [*Technical Communication*]

Suomen tiedetoimittajat (Finnish Association of Science Editors and Journalists). President: Professor Paul Fogelberg, Department of Geography, University of Helsinki, Hallituskatu 11, 00100 Helsinki, Finland; Secretary: Ms Kerttu Tirronen, Technical Research Centre of Finland (VTT), Information Service, Vuorimiehentie 5, 02150 Espoo, Finland

Wetenschappelijke-Redacteurenkring (WERK)/Netherlands Association of Science Editors (NASE). President: Dr Arie A. Manten, Elsevier Science Publishers, P.O. Box 2400, 1000 CK Amsterdam, The Netherlands; Secretary: Ir A.K.S. Polderman, Royal Dutch Association of Pharmacy, Alexanderstraat 11, 2514 JL 's-Gravenhage, The Netherlands

(2) Some organizations and individuals offering courses on editing/copyediting or on publishing

a) In the United States

CUNY Graduate Center (Education in Publishing Program), 33 West 42 St, New York, NY 10036, tel. 212 790–4453/4 [evening seminars or workshops on copyediting/proofreading and line editing]

Denver Publishing Institute, University of Denver, 2075 South University, #D-114, Denver, CO 80210 [a four-week program concentrating on book publishing, with hands-on workshops in editing and many other aspects of publishing; not particularly directed towards science or copyediting]

Anita DeVivo, Consultant, 5480 Wisconsin Avenue, Suite 1630, Chevy Chase, MD 20815, tel. 301 656–3112

Editorial Experts Inc. (see part 1 above)

George Washington University, Publication Specialist Program, Suite 1409, 801 22nd Street NW, Washington, DC 20052

Howard University Press Book Publishing Institute, 2900 Van Ness Street NW, Washington, DC

New York University's Center for Publishing, Summer Institute in Book and Magazine Publishing, The Publishing Institute, New York University, Rm. 21, 2 University Place, New York, NY 10003, tel. 212 598–2101 [three intensive weeks on books publishing, followed by three weeks on magazines; not directed towards copyediting in science]

Northwestern University, College of Continuing Professional Education, 339 E. Chicago Avenue, Chicago, IL 60611

Radcliffe Publishing Procedures Course, 6 Ash Street, Cambridge, MA 02138

Barbara B. Reitt, Reitt Editing, 3505 Hampton Hall Way, NE, Atlanta, GA 30319, tel. 404 255–5790

Rice Publishing Program, Box 1892, Houston, TX 77251

Stanford Publishing Course, Bowman Alumni House, Box WMF, Stanford, CA 94305, tel. 415 497–2021 [offers an intensive 12-day course on book and magazine publishing; not directed towards science or copyediting but good background training]

University of California, Certificate Program in Publishing, University of California Extension, 2223 Fulton Street, Berkeley, CA 94730

University of Chicago

And see what your local educational establishments have to offer.

b) In the United Kingdom

Book House Training Centre (Publishers Association), 45 East Hill, London, SW18 2QE, tel. 01-874 4608 [occasional short courses for desk editors]

London College of Printing, Elephant and Castle, London, SE1, tel. 735 8484 [courses in book and periodical production]

London School of Publishing, 47 Red Lion Street, London, WC1R 4PF, tel. 01-405 9801 [evening and other courses in editing and related skills]

National Extension College, 18 Brooklands Avenue, Cambridge CB2 2HN [correspondence courses]

Oxford Polytechnic, Headington, Oxford, OX3 0BP, tel. 0865 64777 [diploma in publishing]

PMA Publishing Services, Philip Marsh Associates, Belvedere House, High Street, Esher, Surrey, KT10 9LG, tel. 0372 67727 [workshops or seminars, for editorial staff, including copyeditors]

Diplomas or MA courses in publishing may also be available at the following:

Exeter College of Art & Design, Department of Printing, The Mint, Exeter, EX4 3DL

Napier College of Commerce and Technology, Colinton Road, Edinburgh, EH10 5DT

University of Leeds, School of English, Leeds, LS2 9JT

Watford College, Hempstead Road, Watford, WD1 3EZ

Further Reading

Some books cited in Chapters 1–10 also appear below, but see the reference lists after those chapters for other references.

(1) Style manuals and related reference books

American Institute of Physics. 1978   Style manual. American Institute of Physics, 1978.

American Psychological Association. 1983   Publication manual of the American Psychological Association, 3rd ed. American Psychological Association, Washington, DC, 1983.

American Society of Agronomy, Crop Science Society of America, Soil Science Society of America. 1984   Handbook and style manual. American Society of Agronomy, Crop Science Society of America, Soil Science Society of America, 1984.

American Society for Microbiology. 1985   ASM style manual for journals and books. American Society for Microbiology, Washington, DC, 1985.

CBE Style Manual Committee. 1983   CBE style manual: a guide for authors, editors, and publishers in the biological sciences, 5th ed. Council of Biology Editors, Bethesda, MD, 1983.

Cochran W, Fenner P, Hill M (eds.). 1979   Geowriting: a guide to writing, editing, and printing in earth science, 3rd ed. American Geological Institute, Alexandria, VA, 1979.

Dodd JS (ed.). 1986   The ACS style guide: A manual for authors and editors. American Chemical Society, Washington, DC, 1986.

Follett W. 1974   Modern American usage. Warner, New York, 1974.

Fowler HW. 1965   A dictionary of modern English usage, 2nd ed., revised by Sir Ernest Gowers. Oxford University Press, Oxford, 1965.

Glen JW (ed).   1977–   Editerra editors' handbook. EASE, Bromley, Kent
(obtainable from Geo Books, Regency House, 34 Duke Street, Nor-
wich, NR3 3AP, UK), 1977–. (Includes sections on indexing and retrieval, computer programs in science journals, mineralogical nomenclature, rock nomenclature, and stratigraphic nomenclature. New
sections are added from time to time.)

Hart H.   1983   Rules for compositors and readers at the University Press,
Oxford, 39th ed. Oxford University Press, Oxford, 1983.

Howell JB.   1983   Style manuals of the English-speaking world: a guide.
Oryx Press, 2214 North Central at Encanto, Phoenix, AZ 85004.

Miller C, Swift K.   1980   The handbook of non-sexist writing. Lippincott & Crowell, New York, 1980/Women's Press, London, 1981.

Skillin M, Gay R.   1974   Words into type. Prentice-Hall, Englewood Cliffs,
NJ, 1974.

Svartz-Malmberg G, Goldmann R. (eds. for Nordic Publication Committee for Medicine).   1978   Nordic biomedical manuscripts: instructions
& guidelines. Universitetsforlaget, Oslo, 1978.

U.S. Geological Survey.   1986   Suggestions to authors of the reports of
the United States Geological Survey, 7th ed. U.S. Government Printing Office, Washington, DC (in press).

U.S. Government Printing Office.   1986   A manual of style – a guide to
the basics of good writing. Gramercy, New York, 1986.

University of Chicago Press.   1982   The Chicago manual of style, 13th
ed. University of Chicago Press, Chicago, 1982.

(2) Dictionaries

American heritage dictionary of the English language.   1984   Houghton
Mifflin, Boston, 1984.

Butterworth's medical dictionary, 2nd ed. Butterworths, London, 1978.

Concise Oxford dictionary of current English, 7th ed. Oxford University
Press, Oxford, 1982.

Dorland's illustrated medical dictionary, 26th ed. Saunders, Philadelphia,
1981.

International dictionary of medicine and biology. Wiley, New York, 1986.

McGraw-Hill dictionary of scientific and technical terms, 2nd ed. McGraw-
Hill, New York, 1983.

Random House college dictionary, revised ed. Random House, New York,
1982.

Stedman's medical dictionary, 24th ed, illustrated. Williams & Wilkins, Baltimore, 1981.

Webster's third new international dictionary of the English language, unabridged. Merriam, Springfield, MA, 1961.

(3) Nomenclature (see also the manuals listed in Part 1)

Buchanan RE, Gibbons NE (eds.).   1974   Bergey's manual of determinative bacteriology, 8th ed. Williams & Wilkins, Baltimore, 1974.

Marler EEJ (compiler).   1985   Pharmacological and chemical synonyms: a collection of names of drugs, pesticides and other compounds drawn from the medical literature of the world, 8th ed. Elsevier, Amsterdam, 1985.

Nomina Anatomica, 4th ed., together with Nomina Histologica and Nomina Embryologica. Excerpta Medica, Amsterdam, 1977.

Skerman VBD, McGowan V, Sneath PHA (eds.).   1980   Approved lists of bacterial names. American Society for Microbiology, Washington, DC, 1980 [reprinted from the International Journal of Systematic Bacteriology 1980; 30:225–420].

Weast RC (ed.).   1986   Handbook of chemistry and physics, 66th ed. CRC Press, Boca Raton, FL, 1986.

Webb EC (ed.).   1984   Enzyme nomenclature: recommendations (1984) of the Nomenclature Committee of the International Union of Biochemistry. Academic Press, Orlando, FL, 1984.

Willis JC (revised by HK Airy Shaw).   1973   A dictionary of the flowering plants and ferns, 8th ed. Cambridge University Press, Cambridge, 1973.

Windholz M et al (eds.).   1983   The Merck index: an encyclopedia of chemicals, drugs, and biologicals, 10th ed. Merck, Rahway, NJ, 1983.

(4) Books on editing, scientific writing, and related subjects

Bénichoux R, Michel J, Pajaud D et al.   1985   Guide pratique de la communication scientifique: comment écrire, comment dire. Gaston Lachurié, Paris, 1985.

   Highly recommended for scientists writing or lecturing in French.

Bishop C.   1984   How to edit a scientific journal. ISI Press, Philadelphia, 1984.

   Emphasis is on how to select and improve manuscripts.

Booth V.   1985   Communicating in science: writing and speaking. Cambridge University Press, Cambridge, 1985.

   Pithy. Says it all in 80 pages.

Butcher J.   1981   Copy-editing: the Cambridge handbook, 2nd ed. Cambridge University Press, Cambridge, 1981.

   A classic.

Cavendish JM.   1984   A handbook of copyright in British publishing practice, 2nd ed. Cassell, London, 1984.

   A readable survey of a difficult subject.

Cronin B.   1984   The citation process: the role and significance of cita-
tions in scientific communication. Taylor Graham, London, 1984.
   A short (100-page) discussion of citation in its social context.
Day RA.   1983   How to write and publish a scientific paper, 2nd ed. ISI
Press, Philadelphia, 1983.
   Easy-to-read advice for scientists.
DeBakey L.   1976   The scientific journal: editorial policies and
practices — guidelines for editors, reviewers and authors. Mosby, St
Louis, MO, 1976.
   Sound and comprehensive advice, mainly for editors.
Farr AD.   1985   Science writing for beginners. Blackwell Scientific, Ox-
ford, 1985.
   Pleasantly comprehensive; includes a chapter on word processing.
Hall C.   1983   Editing for everyone. National Extension College, Cam-
bridge, UK, 1983.
   For home study.
Huth EJ.   1982   How to write and publish papers in the medical sciences.
ISI Press, Philadelphia, 1982.
   Another comprehensive book of good advice for medical scientists.
Huth EJ.   1986   Medical style and format: an international manual for
authors, editors, and publishers. ISI Press, Philadelphia, 1986.
   Detailed advice on style; invaluable for copyeditors working in all
branches of science, as well as in medicine.
Judd K.   1982   Copyediting: a practical guide. Kaufmann, Los Altos, CA,
1982.
   Very readable and useful guide.
Kachergis J et al.   1977   One book/five ways. William Kaufmann, Los
Altos, CA.
   A work book showing in detail how five different publishers would
have handled the same manuscript. Useful background for anyone
entering publishing.
King LS.   1978   Why not say it clearly: a guide to scientific writing. Lit-
tle, Brown, Boston, 1978.
   Readable guide to prose style for scientists.
Kirkman J.   1980   Good style for scientific and engineering writing. Pit-
man, London, 1980.
   Good practical advice. Includes the results of a survey on the writ-
ing styles preferred by scientists when they are reading.
Lock S.   1977   Thorne's better medical writing, 2nd ed. Pitman Medical,
Tunbridge Wells, 1977.
   Covers the MD thesis, writing for money, papers at meetings, and
much else.
Lock S.   1985   A difficult balance: editorial peer review in medicine.
Nuffield Provincial Hospitals Trust, London, 1985/ISI Press, Philadel-
phia, 1986.

Essential reading for anyone who wants to know more about refereeing/peer review in science (not just medicine).

Morgan P.   1986   An insider's guide for medical authors and editors. ISI Press, Philadelphia, 1986.

Delightful and informative collection of essays on editing and writing.

O'Connor M, Woodford FP.   1975   Writing scientific papers in English: an ELSE–Ciba Foundation guide for authors. Associated Scientific Publishers [Excerpta Medica], Amsterdam, 1975 [paperback Pitman, London, 1978].

Second edition in preparation (due 1987; Wiley, Chichester and New York).

O'Connor M.   1978   Editing scientific books and journals: an ELSE–Ciba Foundation guide for editors. Pitman, London, 1978 [published in the United States as The scientist as editor. Wiley, New York, 1979].

Plotnik A.   1982   The elements of editing: a modern guide for editors and journalists. Macmillan, New York, 1982 (paperback 1984).

Splendidly readable account of editing; not aimed at scientific editors but has much for them too.

Reynolds L, Simmonds D.   1981   Presentation of data in science: publications, slides, posters, overhead projections, tape-slides, television — principles and practices for authors and teachers. Martinus Nijhoff, The Hague, 1981.

Very useful for the do-it-yourself author, or newsletter producer.

Strunk W, White EB.   1978   The elements of style, 3rd ed. Macmillan, New York, 1978.

Short, simple, superb, and essential for anyone needing advice on grammar and style.

Tichy HJ.   1966   Effective writing for engineers, managers, scientists. Wiley, New York, 1966.

A very thorough guide to writing.

Woodford FP (ed.).   1986   Scientific writing for graduate students: a manual on the teaching of scientific writing, 4th printing. Council of Biology Editors, Bethesda, MD, 1986.

Another classic: intended for teachers of scientific writing but equally useful for the writers themselves.

(5) Articles on copyediting: a short list

Bostian LR.   1986   Working with writers. Scholarly Publishing 1986; 17:119–126.

Broadbent M.   1979   Productivity in copy editing. Scholarly Publishing 1979; 10:170–174.

Broadbent M.   1985   Checklist for copy editors. CBE Views 1985; 8(1):36–38.

Brogan M. 1979   Costs in copy editing. Scholarly Publishing 1979; 10:47–53.

Bryson V, Nemiah JC, Frost D.   1980   Author's editors: three comments. CBE Views 1980; 3(2):14–18.

Burgan MW.   1984   Consistency in nomenclature. CBE Views 1984; 7(3):12–14.

Core G.   1974   Costs and copy-editing. Scholarly Publishing 1974; 6:59–65.

Evans N.   1979   Line editors: the rigorous pursuit of perfection. Publishers Weekly, 15 October, p. 24–31.

Harman E.   1975   Hints on poofreading. Scholarly Publishing 1975; 6:151–157.

Hundley B.   1982   Manuscripts on tape. CBE Views 1982; 5(2):6–12.

Pascal NB.   1982   How much editing is enough? Scholarly Publishing 1982; 13:263–268.

Stainton EM.   1977   A bag for editors. Scholarly Publishing 1977; 8:111–119.

Stainton EM.   1977   A bag for authors. Scholarly Publishing 1977; 8:335–345.

Stainton EM.   1978   A mixed bag: getting along together. Scholarly Publishing 1978; 9:149–158.

Stainton EM.   1978   Another mixed bag. Scholarly Publishing 1978; 9:219–230.

Stainton EM.   1985   The copy editor. Scholarly Publishing 1985; 17:55–63.

Tacker MM.   1980   Author's editors: catalysts of scientific publishing. CBE Views 1980; 3(1):3–11.

(6) Periodicals

See the publications of the associations listed in the appendix, part 1. The book trade weeklies, *Publishers Weekly* (Bowker, New York) and *The Bookseller* (J. Whitaker, London), are useful for keeping in touch with what's going on in publishing in general; *Science, Nature,* and *New Scientist* are useful for keeping in touch with what's going on in science.

Index